Contents

Lecture 1 3

An introduction to telephony System 3

   1.1   **Introduction** 3

   1.2   Basic telephony 6

   1.3   A telephone networks 20

   1.4   How does a network set up a call connection? 28

   1.5   Waveforms 30

   1.6   Other forms of telephone networks 36

         Definition: Telegraph (from Greek): 48

Lecture 2 51

Voice Transmission 51

   2.1 Basic Concepts 51

   2.2 Pulse Amplitude Modulation 61

   2.3 Pulse Coded Modulation (PCM): 64

   2.4 2 Mbit/s Frame and Signaling Pulse Frame 78

   2.5 T1 Carrier System: 81

   2.5 Baseband Transmission of Digital Signals and codic 83

   2.6 Digital Transmission (Line Coding): 97

TDM and Codecs 117

Lecture 3 118

Switching 118

   3.1   Introduction 118

   3.2 Crossbar Switch 118

   3.3 Space Division Switching 120

3.4 Nonblocking Switching: 121

3.5 Three-Stage Switch Design 123

3.6 Implementation Complexity of TDS 126

Lecture 4 130

Backbone Transmission 130

    4.1   Transmission Media 130

    4.2   Digital Transmission systems 137

    4.3   Wave Division Multiplexing (WDM): 153

Lecture 5 210

Access Networks 210

    5.1 ADSL Modems 210

    5.2 Radio Wave Propagation 216

    5.2 GPON –Fundamentals 227

References 233

Lecture 1
An introduction to telephony System

1.1 Introduction

Telecommunications is today widely understood to mean the electrical means of communicating over a distance. The first form of telecommunications was that of the Telegraph, which was invented quite independently in 1837 by two scientists, Wheatstone and Morse. Telegraphy was on a point-to-point unidirectional basis and relied on trained operators to interpret between the spoken or written word and the special signals sent over the telegraph wire. However, the use of telegraphy did greatly enhance the operations of railways and, of course, the dissemination of news and personal messages between towns. This usefulness of telecommunications on the one hand and the limitation of needing trained operators on the other led to the aspiration for a simple means of bi-directional voice telecommunications that anyone could use. Alexander Graham Bell met this need when he invented the telephone in 1876. Remarkably soon afterwards, the World's first telephone exchange was opened in 1878 in New Haven, Connecticut, USA. Since then, telephony has become the ubiquitous means of communicating for humankind, and telephone networks using the principles of Alexander Graham Bell have been implemented throughout the World. This chapter introduces the basic principles of telephony, covering the operation of a telephone and the way that telephones are connected via a network.

- Telecommunication means "communications at a distance" (Tele in Greek means at a distance)
- Electrical communications by wire, radio, or light (fiber optics)
- Traditionally two distinct disciplines:

> Switching: selects and directs communication signals to a specific user or a group of users

> Transmission: delivers the signals in some way from source to the far-end user with an acceptable signal quality

- The source may be a simple telephone microphone, keyboard
- The destination may be a simple telephone speaker, monitor

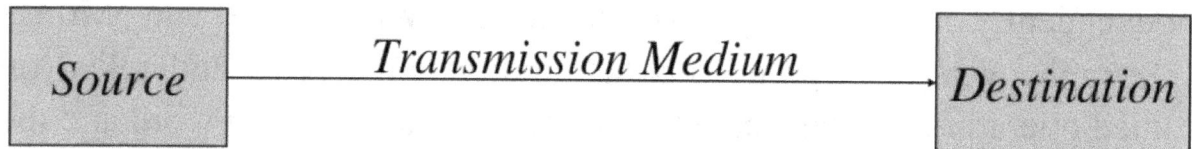

- It can be a seen as a single electrical medium

- Or, as a cascade of electrical media

- Networks show a gain or loss.
- To understand these gains or loss, a good knowledge of the decibel and related measurement units is needed.
- the dB (decibel) is a relative unit of measurement commonly used in communications for providing a reference for input and output levels.

  > Power gain or loss.

- Decibels are used to specify measured and calculated values in

  > audio systems, microwave system gain calculations, satellite system link-budget analysis, antenna power gain, light-budget

calculations and in many other communication system measurements

> In each case the dB value is calculated with respect to a standard or specified reference.

- The dB value is calculated by taking the log of the ratio of the measured or calculated power (P2) with respect to a reference power (P1).

$$P_1 \longrightarrow \boxed{} \longrightarrow P_2$$

- The result is multiplied by 10 to obtain the value in dB.

$$dB = 10 \log_{10} \frac{P_2}{P_1}$$

- It can be modified to provide a dB value based on the ratio of two voltages. By using the power relationship $P = V^2/R$

$$dB = 10 \log_{10} \frac{P_2}{P_1} = 10 \log_{10} \frac{V_2^2/R}{V_1^2/R} = 20 \log_{10} \frac{V_2}{V_1}$$

- dBm indicates that the specified dB level is relative to a 1-milliwatt reference.

$$1\,mW \longrightarrow \boxed{} \longrightarrow P_2$$

$$dBm = 10 \log_{10} \frac{P_2}{0.001\,W}$$

- If Power is expressed in watts instead of milliwatts.

 the dB unit is obtained with respect to 1 watt and the dB values are expressed as Dbw

$$dBW = 10 \log_{10} \frac{P_2}{1\,W}$$

Examples:

- Important Note: The decibel (dB) is "the logarithm of a power ratio" and NOT a unit of power;
- However, dBW and dBm are units of power in the logarithmic system of numbers
  - Convert the following into dBm or dBW
  - P=1mW, P(dBm)=?
  - P=0.1mW, P(dBm)=?
  - P=10W, P(dBW)=?
  - P=1W, P(dBm)=?

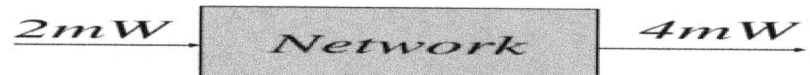

- dB value=10log104/2=10log102=10×0.3010=+3.01dB≈3dB
  - Memorize the above relationship
  - The amplifying network has a 3-dB gain because the output
  - power was the double the input power

1.2 Basic telephony

Definition: Telephone (from Greek):
- tele- means "afar, far off"
- -phone means "sound, voice"
  - Telephone is an instrument that transmits sound, voice to a distant place

Before the invention of telephone, there was telegraph which was telephone's precedent.

- In the early 1870s, while Alexander Graham Bell was experimenting with telegraph, he realized that transmission the human voice over a wire by using electricity might be possible.
- Because he lacked the skill to make the equipment that were necessary for the experiments, he asked Thomas A. Watson for assistance.
- Bell and Watson worked together on the experience which tried to produce sounds over the "harmonic telegraph".
- On June 2, 1875, Bell heard the sound coming to him over the wire.
- After much tinkering, the instrument successfully transmitted the recognizable voice sound, but not words.
- Thus, Bell and Watson spent the whole summer to experiment.
- On March 7, 1876, the patent for telephone was issued to him.
- At 109 Court Street, Boston, the first understandable sentence was carried by the telephone.

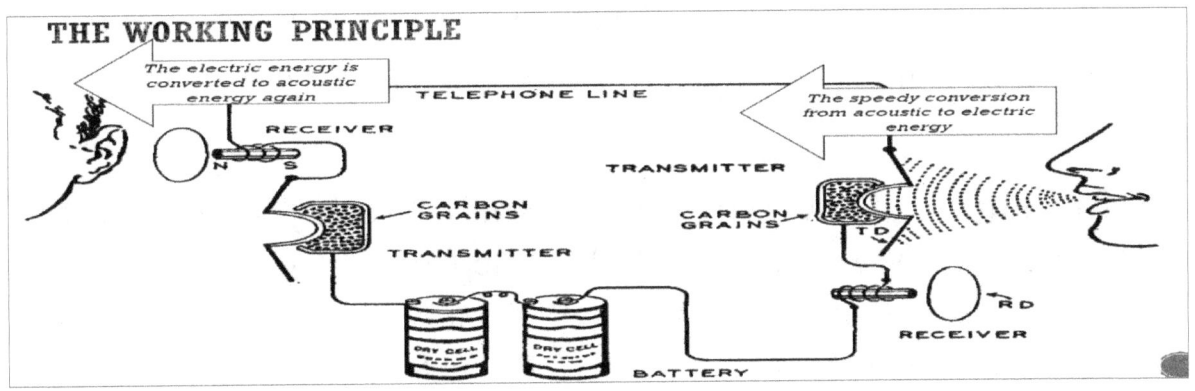

THE PROCESS OF DEVELOPMENT

The first telephone   *The next telephones*

- At first, the telephone was extremely hard for anyone to afford because of its price. Only the rich could afford for telephone.

- Telephone's invention contributed to the development of city centers and buildings.

- With the invention of telephone, some jobs suddenly became unnecessary.

- In addition, the world became much smaller and more accessible to all business because of the telephone.

- In 1885, American Telegraph and Telephone Company (AT&T) was formed.

- The growth of telephone was extremely fast. (By 1900 there were nearly 600,000 phones in Bell's telephone system; that number raised to 2.2 million phones by 1905, and 5.8 million by 1910). (30 millionth phone was connected in the U.S. by 1984, by the 1960s, there were more than 80 million phones in the U.S. and 160 million in the world).

Time	Quantity of telephones (unit)
May 1877	6
November 1877	3,000
1881	133,000
1990	600,000
1905	2,2000,000
1910	5,800,000

The statistic about the growth of telephone

Fig. 1.1(a) illustrates a basic simple one-way telephone circuit between two people. The set-up comprises a microphone associated with the speaker, which is connected via an electrical circuit with a receiver at the remote end associated with the listener. A battery provides power for the operation of the microphone and receiver. During talking, variations in air pressure are generated by the vocal tract of the speaker. These variations in air pressure, known as sound waves, travel from the speaker to the microphone, which converts them into an electrical signal varying in sympathy with the pattern of the sound waves. Indeed, if you are to look at the electrical signal on the circuit leaving the microphone, as illustrated in Fig. 1.1(a), the level of the electrical signal varies with time, with an average value set by the voltage of the battery and with modulations above and below this level, representing the variation in sound pressure hitting the microphone. This electrical signal is an analogue signal because it is an analogue of the sound wave variations in air pressure.

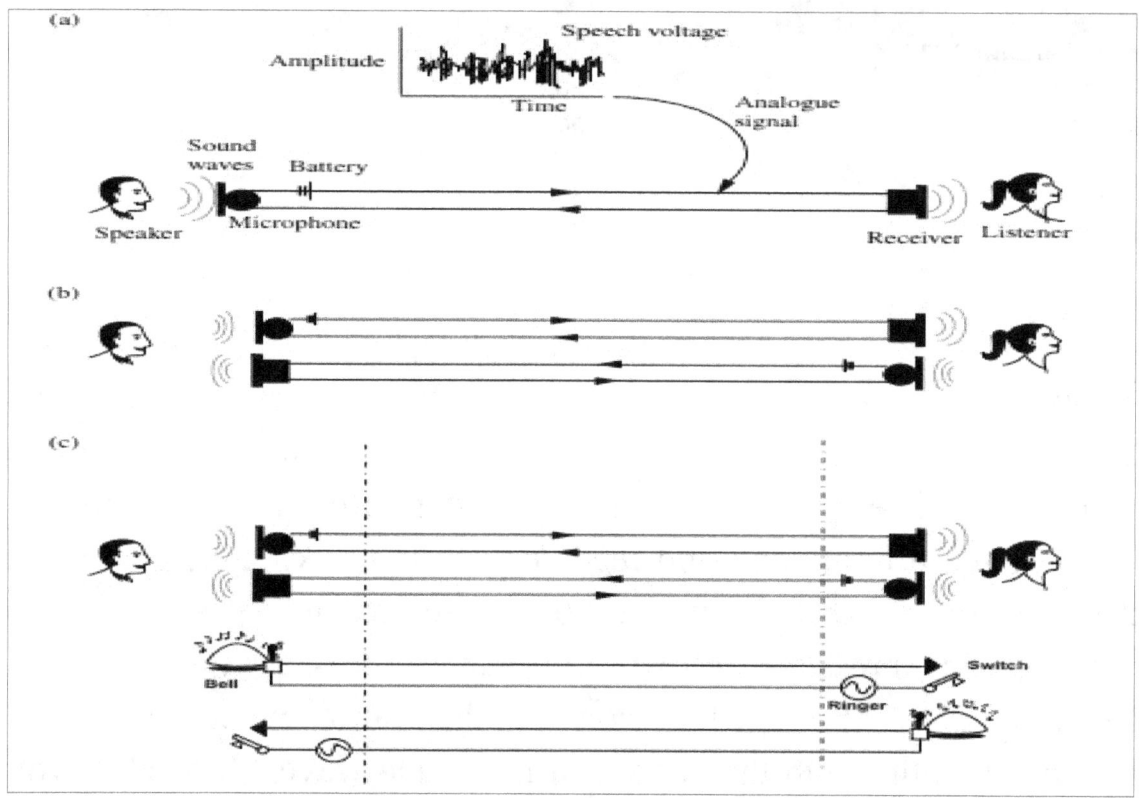

Figure 1.1 (a) Simple One-Way Speech Over Two Wires [Ward]. (b) Both-Way Speech Over Four Wires [Ward]. (c) Both-Way Speech and Alerting Over Eight Wires [Ward]

- The telephone is connected to Public switched
- telecommunications network (PSTN) for local, national, and international voice communications
- The same connections can carry data and image information (television)
- The connection to the PSTN may be via local exchange carriers (LEC)
- End-users, nodes, and connectivity

Voice Telephony:

- Transmission of the human voice

- Voice is a sound signal
- Analog voice-band channel
  - A channel that is suitable for transmission of speech or analog data and has the maximum usable frequency range of 300 to 3400 Hz.
  - The local serving switch is the point of the connectivity with the PSTN
  - It is the point where the analog signal is digitized.

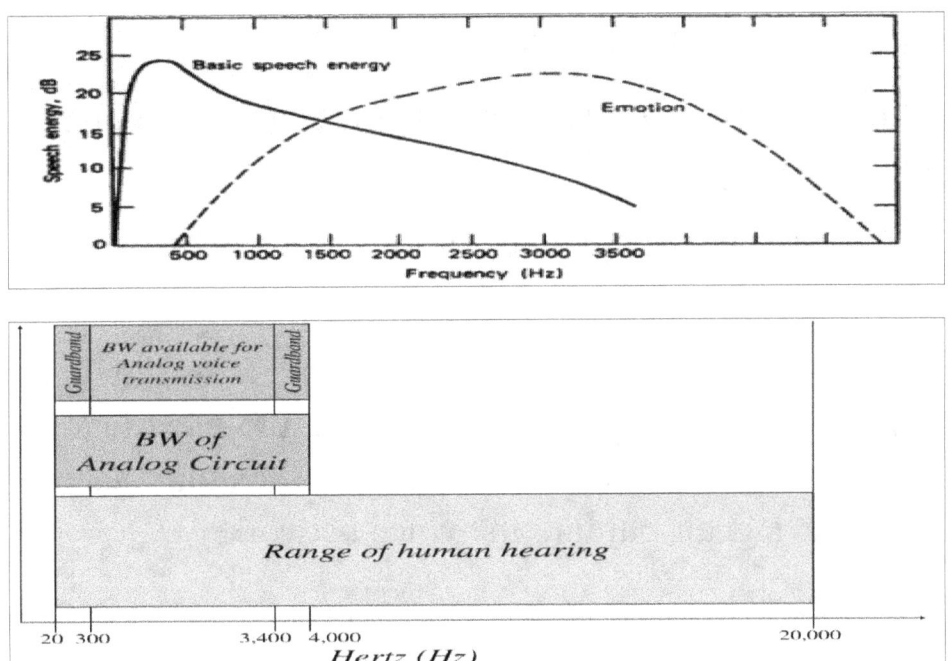

Telephone Subset

- It is a device which converts human speech in the form of sound waves produced by the vocal cord to electrical signals. These signals are then transmitted over telephone wires and then converted back to sound waves for human ears.
- Microphone
- Earphone
- Signaling functions

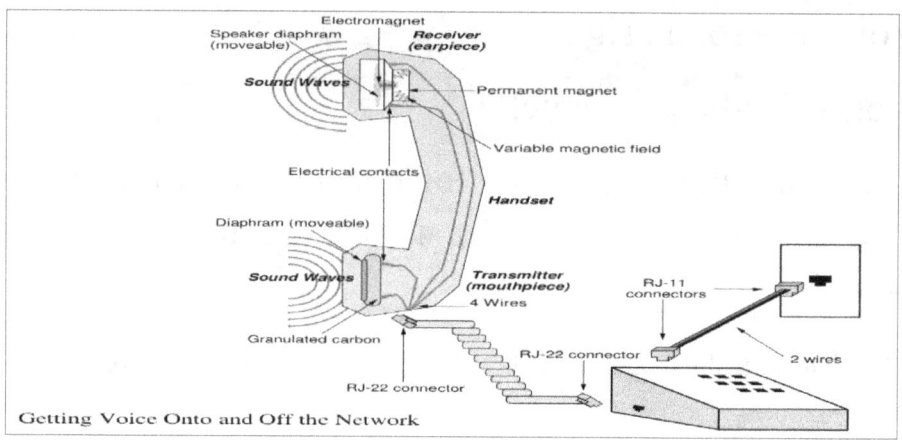

Getting Voice Onto and Off the Network

Telephone Handset

- Microphone (mouthpiece) consists of

  ➢ a movable speaker diaphragm that is sensitive to both amplitude and frequency

  ➢ The diaphragm contains carbon particles that can conduct electricity.

  ➢ As the human voice spoken into the transmitter varies, the amount of carbon granules that strike the electrical contacts in the mouthpiece also varies—thereby sending varying analog electrical signals out into the voice network.

Earphone (earpiece)

➢ Acts in an opposite direction to the mouthpiece.

➢ The electrical signal/waves produced by the transmitter are received at an electromagnet in the receiver.

➢ Varying levels of electricity produce varying levels of magnetism—that, in turn, cause the diaphragm to move in direct proportion to the magnetic variance.

➢ The moving diaphragm produces varying sound that corresponds to the sound waves that were input at the transmitter.

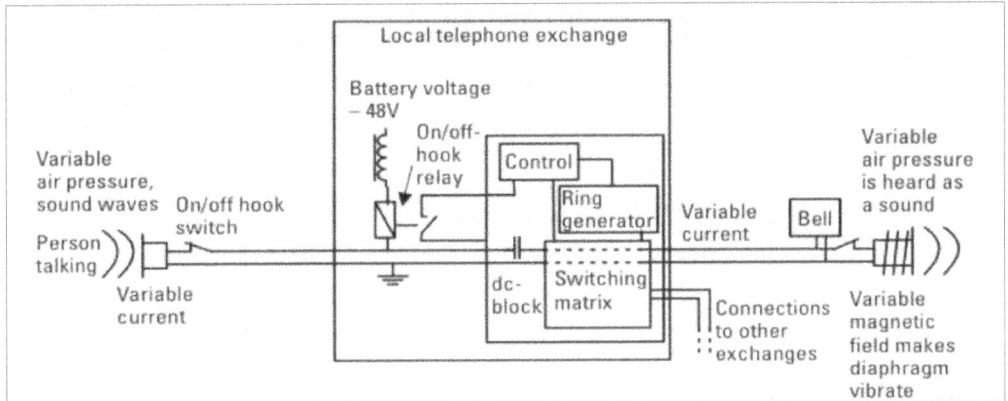

Dialing:

- A combination of 350 Hz and 440 Hz sine waves sent to the Telephone from the central office (CO) indicating that the network is ready to receive calling instructions
- Dialing Modes: Pulse and Touch Tone or Dual-Tone Multi Frequency
- Each button sends a dual frequency sine wave indicated by the corresponding row and column.
- Telephone Numbers are decided by ITU internationally and NANP in North America [NP – numbering plan]
- Pulse
- sends a pulse per digit
- collected by central office
- Tone
- key press (feep) sends a pair of tones = digit
- also called Dual Tone Multifrequency (DTMF)

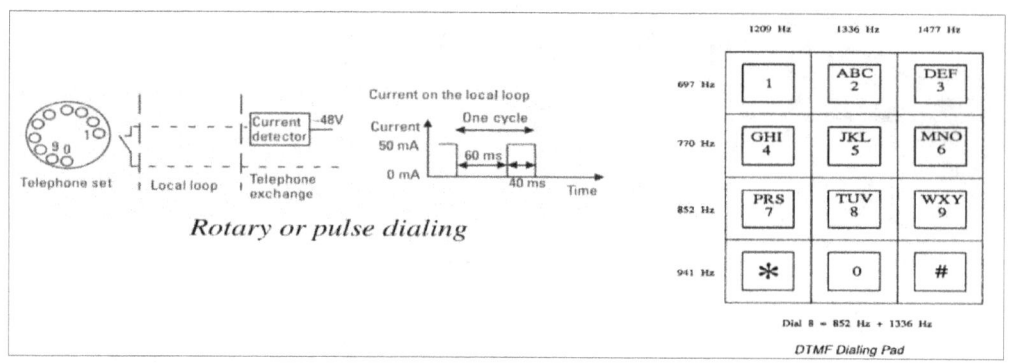

Subscriber Signaling:

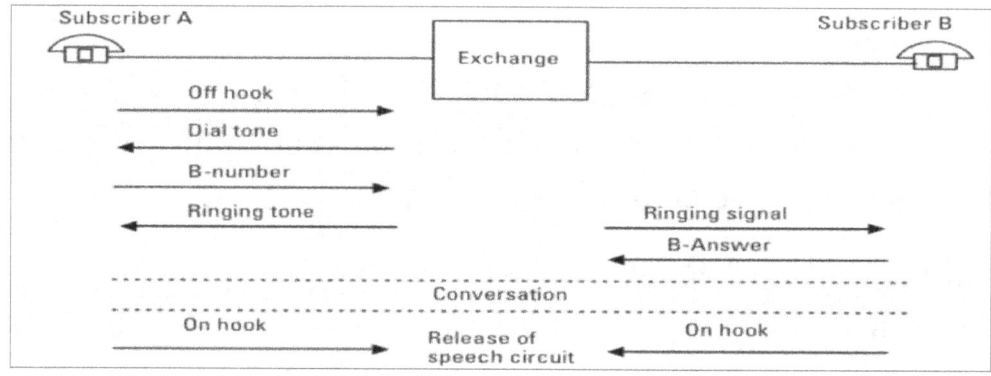

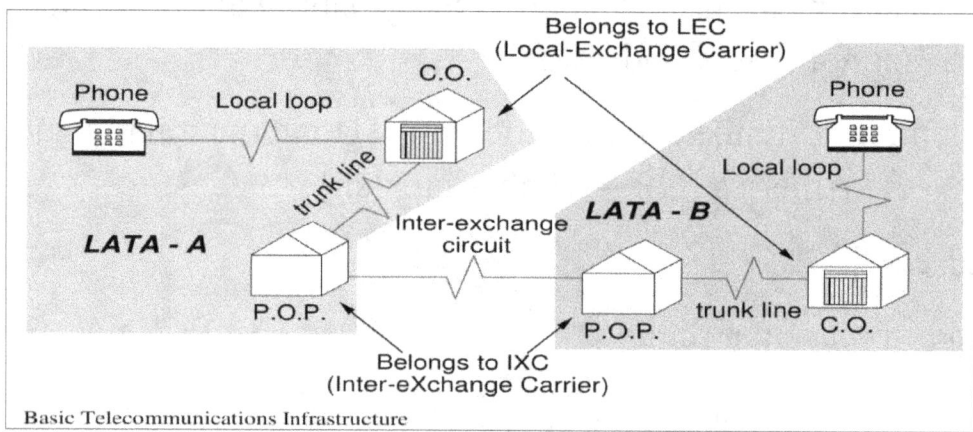

Note 1 Electrical Circuits: Consider an electrical circuit comprising a power source, e.g. a battery, and a length of wire linking both terminals of the power source to some device, say a lamp. Whilst the circuit is complete the lamp will glow and so a switch is normally inserted in to the arrangement to control the light on and off. In this simple example the voltage applied by the battery can be viewed as forcing an electric current to flow around the circuit from its positive terminal to its negative terminal. This flow is referred to a 'direct current' or 'DC'. The lamp contains a coil of special wire that provides an obstacle to the flow of current – known as 'resistance'. The greater the resistance of the lamp the less current the battery can force to flow through the whole circuit. This gives rise to the simple relationship, known as 'Ohm's Law' in which the resistance (measured in ohms or) is given by the voltage (measured in volts or V)

divided by the current (measured in amps or A). An alternative form of electrical voltage is one which cyclically varies from zero up to a maximum positive value, drops to zero and then goes to an equal but opposite maximum negative value and then back to zero. The shape corresponds to the sinusoidal waveform. This so-called alternating voltage creates a corresponding 'alternating current', AC. The electrical main supply is typically at 240 V alternating current (240 V AC), with the cycles occurring at 50 times per second (50 Hz) in the United Kingdom and Europe and at 120 V AC cycling at 60 Hz in the United States. The continuous cycling of the alternating electrical current causes additional changes to the flow of electricity when passing through a circuit. The first phenomenon – capacitance – causes the waveform to be delayed; the second phenomenon – inductance – causes the waveform to be advanced. The results of these effects, known collectively as impedance, are that the AC current is out of step with the applied AC voltage. These effects are used throughout telecommunications and electronic equipment, for example: inductance forms the basic mechanism exploited in hybrid transformers and loading coils.

At the receiving end of the circuit the analogue electrical signal energizes the receiver (i.e. an earpiece), generating a set of sound waves, which are an approximate reproduction of the sound of the speaker.

Obviously for conversation to be possible it is necessary to have transmission in both directions, and therefore a second circuit operating in the opposite direction is required, as shown in Fig. 1.1(b). Thus, a basic telephone circuit comprises four wires: one pair for each speech direction. This is known as a basic 4-wire circuit. In practice, of course, a telephone system would need to include a mechanism for the caller to indicate to the recipient that they wished to speak. Therefore, we need to add to the assembly in Fig. 1.1.(b) a bell associated in a circuit with a power source and a switch. The electrical current flowing in a circuit used to ring a bell in a telephone is known as ringing current. Again, one such arrangement is required in each direction. This argument brings us to the conclusion, illustrated in Fig. 1.1(c), that a set of eight wires, four pairs, is needed to provide bi-directional telephony service between two people. In a practical telephone network, the most important requirement is to minimize the amount of cost associated with connecting each customer. Since there are many thousands or millions of customers on a telephone network, any reduction in the amount of equipment needed to be provided for each customer would result in large overall cost savings.

Thus, some ingenious engineering has enabled significant economy to be achieved through the reduction of the numbers of wires from eight to two, i.e. one pair. This is achieved through the use of 4-to-2-wire conversion (and vice versa), and the time-sharing of functions, as described below.

- 4-to-2-Wire conversion. The two directions of speech circuits, shown in Fig. 1.1(c), can be reduced down to a single circuit carrying speech currents in both directions, using a device known as a hybrid transformer, as shown in Fig. 1.2. (See Box 1.3 for a brief explanation of how a hybrid transformer works.)

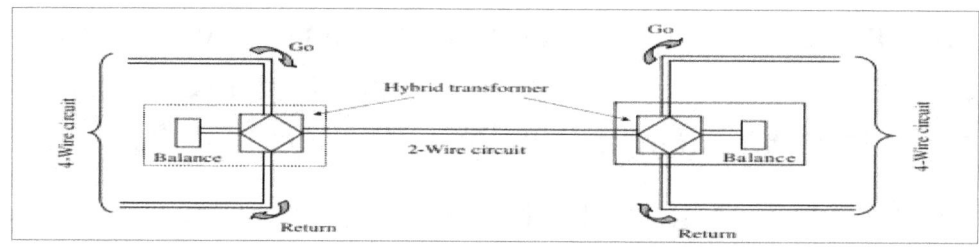

Figure 1.2 4-to-2 Wire Conversion

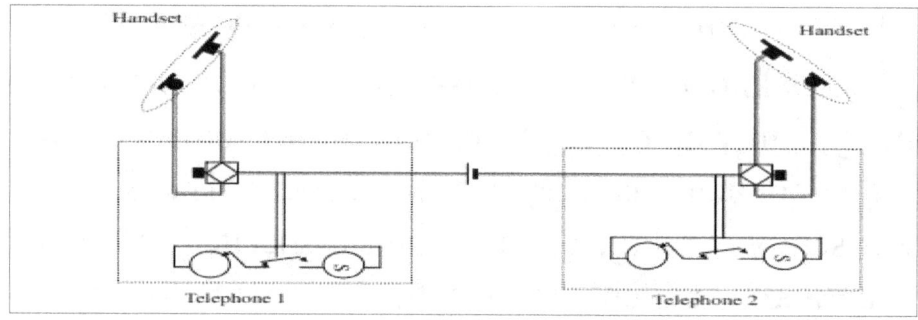

Figure 1.3 A Simple Two-Phone System

- Time-sharing of functions. The need for two pairs of wires to be dedicated to ringing circuits can be totally eliminated by exploiting the fact that ringing does not occur during the speaking phase of a telephone call. Therefore, the single pair provided for speech can instead be used at the start of a call to carry ringing current in either direction, as necessary. Once the call is answered, of course, the single pair is used only to carry the two directions of speech current.

Fig. 1.3 shows that, for our simple two-person scenario, the telephone instrument at each end needs to comprise a handset with microphone and receiver; a hybrid transformer; a bell and a means to send ringing current to the far end. The two telephones need to be connected by a single pair of wires and a battery.

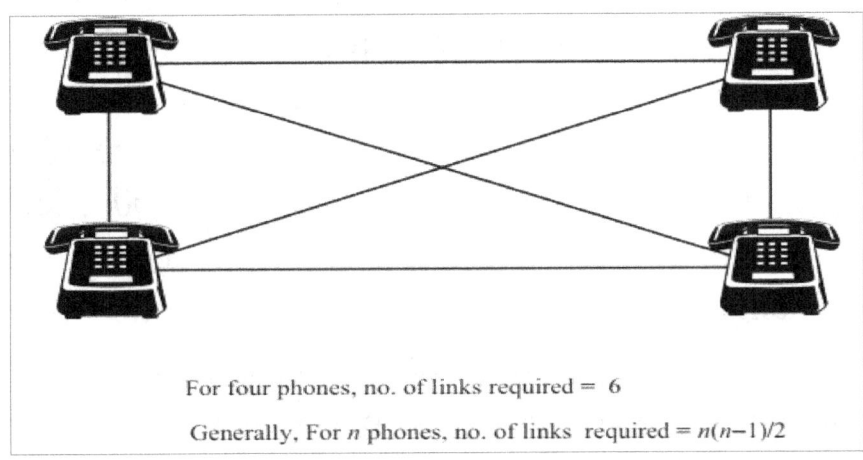

Figure 1.4 Direct Interconnection of Several Phones

We can now extend this basic two-person scenario to the more general case of several people with phones wishing to be able to talk to each other. For example, the logical extension to a four-telephone scenario is shown in Fig. 1.4. In this case six links (i.e. 2-wire circuits) are required in total, each telephone terminating three links. Not only does each telephone need to terminate three links rather than just one, but it also needs a 1-to-3 selection mechanism to choose which of the links should be connected to in order to converse with the required telephone. (Not only does this involve a selection switch within each phone, but the arrangement also needs each phone to be designated with a name or number – as discussed later in Chapter 10.) Whilst the arrangement shown in Fig. 1.4 is quite practical for networks of just a few phones – indeed, many small office and household telephone systems are based on such designs – it does not scale up well. In general, the number of links to fully interconnect n telephones is given by $n(n-1)/2$. As the number of phones becomes large, the number of directly connected links approaches $n2/2$. Clearly, providing the necessary 5,000 direct links in a system serving just 100 telephones would not be an economical or practical design! (In addition, the complexity of the selection mechanism in each telephone would increase in order for it to be capable of switching 1-out-of-99 lines.) The solution to the scaling problem is to introduce a central hub – commonly called an exchange or central office – onto which each phone is linked directly, and which can provide connectivity between any two phone lines, on demand (Fig. 1.5). With a single exchange serving n phones only n links are required; a good solution, which in practice scales up to about 50,000 telephone lines with modern telephone exchanges. We can now deduce the role of a telephone exchange. Fig. 1.6 gives a block schematic diagram of the basic functions required to connect two exchange lines. In the example shown it is assumed that telephone A is calling telephone B. The first requirement is that both telephones A and B need to have an appropriate power source. Although a battery could be provided inside each telephone, indeed in the early days of telephony this was in fact done, it is far more practical to locate the battery centrally at the exchange, where the telephone company (usually known as a Telco') can maintain it. When telephones are connected to their exchange by a pair of metallic wires, usually made of copper, the power for the phones can conveniently be passed over that pair. This arrangement is convenient for the telephone user because then they have no need to manage the charging of batteries in their premises and are also not dependent on the reliability of the local electricity supply. (Although more recently, of course, the need for users to charge the battery in a mobile phone every few days has become acceptable.) However, there are situations where power cannot be passed to the telephone from the exchange. For example, this is not possible when optical fiber is used to connect telephones because glass does not conduct electricity! The other notable example is

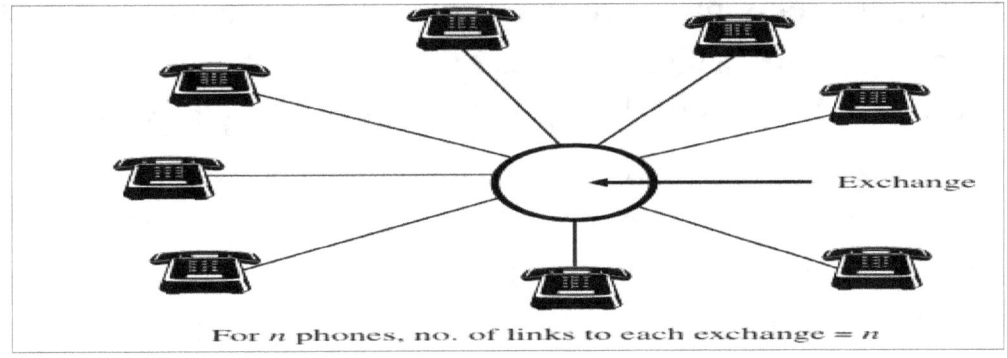

Figure 1.5 Interconnection: Use of an Exchange

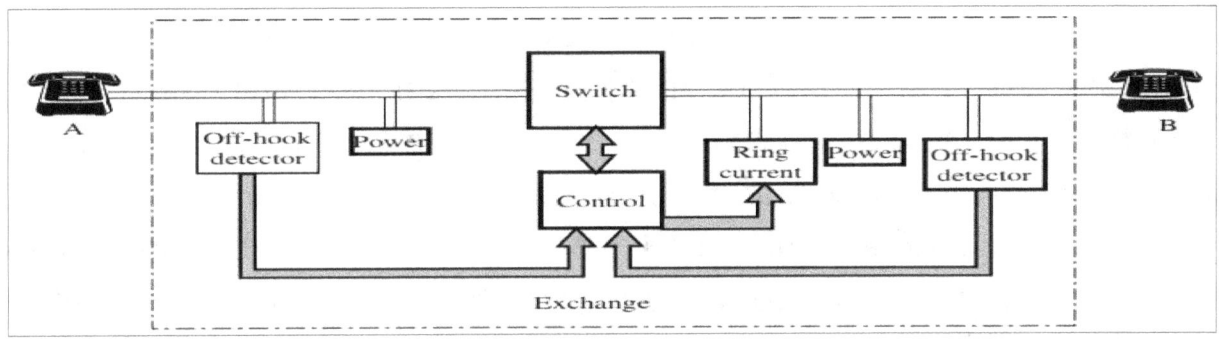

Figure 1.6 The Functions of a Telephone Exchange

that of a mobile network, where a radio path is used to link telephones to the exchange. The first step the exchange has to undertake in managing a call is to detect that the calling telephone (i.e. telephone A in Fig. 1.6) wishes to make a call. The simplest method for conveying such an indication from a telephone, and the one that is still most commonly used today, is for advantage to be taken of the fact that the pair of wires between the exchange and the telephone can be closed at the telephone end, thus creating a loop. This looping of the pair by the telephone causes a current to flow, which operates a relay at the exchange. (A relay is a device that when activated by an electrical current flowing through its coil causes one or more switches to close. The latter in turn then pass electrical current on to other circuits or devices – hence the term 'relay'.) In the case of the original manual exchanges, the closure of the relay switch caused a lamp to glow, hence alerting a human operator to the calling state of the line. For a modern electronic automatic exchange, the energizing of the relay by the loop current causes changes in the state in an electronic system, which is subsequently detected by the exchange control system. The loop is closed in the telephone by a switch activated by the lifting of the handset off the telephone casing – this causes a small set of hooks to spring up. The lifting of the handset creates a condition known as 'off-hook'. In the case of a cordless phone this off-hook switch is located in the base unit (attached to the copper pair termination in the house) and is controlled remotely from the radio handset when the subscriber presses the 'dial' or 'send' button – often indicated on the button by a picture of a handset being lifted. On the outgoing side of the exchange attached to the line for telephone B there is a power source, a generator to send ringing current and a relay to detect the 'off-hook' condition when the called telephone answers. (For simplicity, Fig. 1.6 assumes that telephone A is calling telephone B so the ringing current generator is shown attached only to the line B, but of course any telephone can initiate calls, so in practice all lines have power supply, off-hook detector (i.e. relay) and a ringing current generator.) When telephone exchanges were first introduced the method of connecting two telephone lines together was through a human operator using a short length of a pair of wires across a patching panel. Each telephone line terminated on the panel in a socket with an associated small indicator lamp. The human operator was made aware that telephone A wished to make a call by the glowing of the relevant lamp, activated by telephone A going 'off hook'. On seeing the glowing lamp, the operator started the procedure for controlling a call by first talking to caller A and asking them to which number they wished to be connected. The operator then checked the lamp associated with the called line, if this was not glowing then that line was free, and the call could be established. The next step was for a ringing current to be applied to telephone B. The operator did this by plugging a line connected to a special ringing generator into the socket for telephone B. It was important for the operator to monitor B's lamp to ensure that the ringing was stopped as soon as B answered – otherwise there would be a very annoyed person at the end of the line! The operator would then make the appropriate connection using a jumper wire across the patching panel. Finally, the operator needed to monitor the two lamps involved in the call so that the connection could be taken down (by removing the connecting cord between the two sockets) as soon as one of the lamps went out. The call had then been terminated. In making a call connection, the operator had to follow certain procedures, including writing on a ticket the number of the caller and called lines, the time of day and the duration of the call, so that a charge could be raised later. It is important to remember that an exchange needs to serve many lines and that at any time there will be several calls that need to be set up, monitored or cleared down. The human operators had to share their attention across many calls; each operator typically was expected to be able to deal with up to six calls simultaneously. Generally, today telephone exchanges are fully automatic. However, there are still occasions when a human intervention is required, e.g. providing various forms of assistance and emergency calls, and special auto-manual exchanges with operators provide such services. For convenience, the telephone exchanges considered in the remainder of this book will be only the fully automatic types. The description above is based on a fully manual exchange system because it enables the simple principles of call connection to be explained in a low-technology way – yet, all of the steps and the principals involved are followed in automatic exchange working. In an automatic exchange, as shown in Fig. 1.6, the role of the operator is taken by the exchange control (representing the operator's intelligence), which in modern exchanges is provided by computers with the procedures captured in call-control software, and the switch or 'switch block', usually in the form of a semi-conductor matrix, which performs the functions of the connecting cords and patch panels.

1.3 A telephone networks

A telephone exchange serves many telephone lines, enabling any line to be connected to any other (when they are both free). In a small village all the lines could easily be connected to a single telephone exchange, since the distances are short. However, if telephone service needs to be provided to a larger area, the question arises as to how many exchanges are needed. This is illustrated in Fig. 1.7, where a region of the country has a large population of telephones to be served. They could all be served by one central large exchange or by several smaller exchanges. Obviously, the lengths and hence costs of the telephone lines reduce as the number of exchanges increases, but this saving is offset by the increase in costs of exchange equipment and buildings. In addition, a link (known as a 'junction route') needs to be established between each exchange to ensure full connectivity between all telephones in the region. The trade-off between the cost of the telephone lines and the costs of the switching and buildings and junction-route costs plotted for various numbers of exchanges, n, to serve the population of telephones in the region shows a typical bath-curve shape [3]. In this example of Fig. 1.7, the optimum cost is achieved with three exchanges. Of course, it is not only the number of exchanges that contributes to the optimum costs, but also the location of the exchanges within their catchment areas. The optimum total cost for the region is achieved when the exchanges are at the center

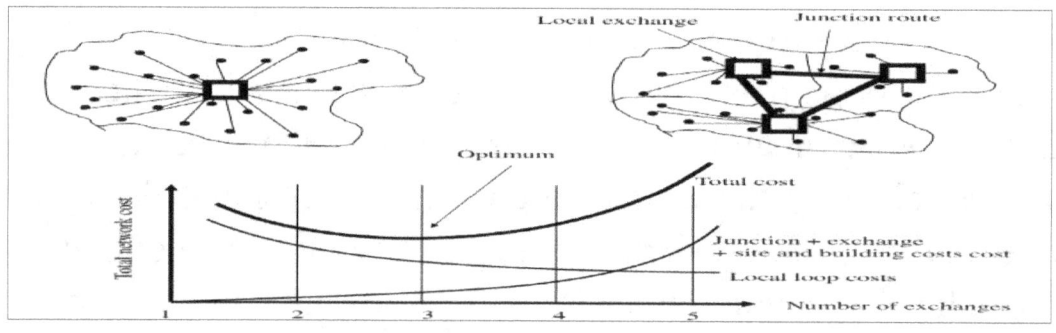

Figure 1.7 Network Optimization [Ward]

of gravity (i.e. the location where the sum of all the line lengths is the minimum) of the population of telephones served. In practice, network operators locate telephone exchanges as close to this center of gravity as possible, within the constraints of the availability of suitable sites within a town. There are also practical limitations on the lengths of telephone lines which constrain the size of the catchment area of lines dependent on one exchange. These limits are set by the electrical characteristics of the lines, predominantly the resistance of the loop. (See Box 1.1 for an explanation of resistance.) Typically, this resistance needs to be less than 2,000 ohms (written as '2000') to ensure that sufficient current flows for the 'off hook' condition to be detected by the exchange and also to ensure that the loudness of the call is acceptable. There are several ways in which the telephone lines can be kept within the electrical limits, including the use of thicker gauge wire (more expensive, but having less resistance) on the longer telephone lines. Also, in the United States, where the terrain requires larger catchment areas, inductors (i.e. devices comprising tightly coiled wire), known as 'loading coils', are added to long lines to reduce the signal loss. The majority of telephone lines in the United Kingdom are below 5 km, whereas in the United States there are many lines in excess of 10 km and exchange catchment areas can be as large as 130 square miles [4]. These aspects are considered in more detail in Chapter 5. Thus, the primary design requirement for a network operator is to achieve a cost optimized set of exchanges, each of which is located at the center of gravity

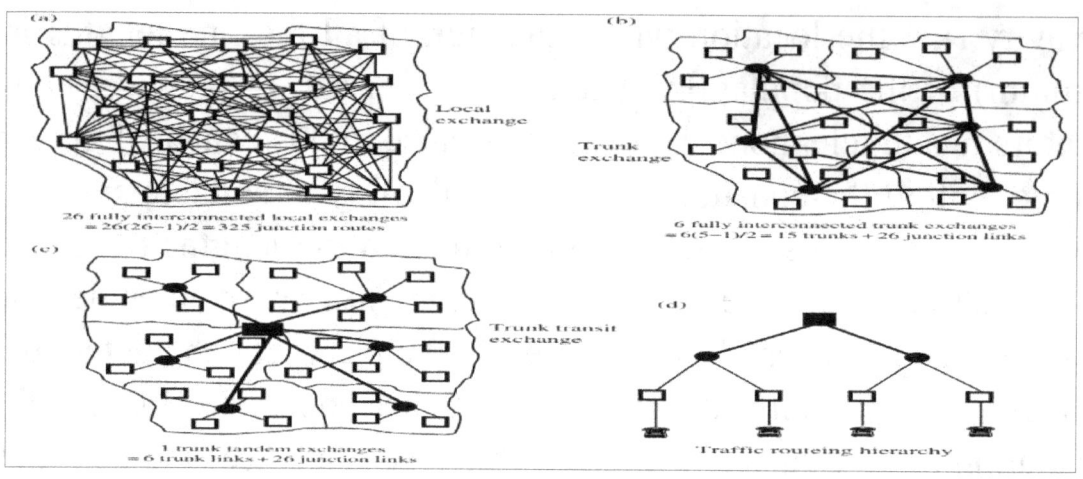

Figure 1.8 (a) Network Structure-1 [Ward]. (b) Network Structure-2 [Ward].(c) Network Structure-3 [Ward]. (d) Network Structure-4 [Ward]

for the catchment area, whilst keeping within the electrical limits on telephone line lengths. We can now consider extending this logic to a telephone network serving several regions, or even the entire country. Given that each exchange needs to be able to connect to every other exchange (via junction routes), a further network optimization issue arises. For example, for an area comprising 26 exchanges the number of fully interconnected junction routes would be 26(26−1)/2 = 325, as shown in Fig. 1.8(a). (For a country like the United Kingdom with some 6,000 local exchanges (LEs) the number of directly connected junction routes would be an impossible 17,997,000!).As before, this situation is alleviated by the use of a central exchange within each region – known as a trunk exchange (TE). These trunk exchanges also need to have links (i.e. 'trunk routes') between them to ensure full connectivity across all regions. In the example of Fig. 1.8(b) this arrangement requires just 15 trunk links and 26 junction links, but with the added cost of 6 trunk exchanges. Further network optimisation is possible by the addition of a single trunk tandem exchange to provide connectivity between each of the trunk exchanges, reducing the number of trunk links to six, but with the added cost of a trunk tandem exchange (see Fig. 1.8(c)). The arrangement of exchanges and links can be drawn as a traffic routeing1 hierarchy, as shown in Fig. 1.8(d). (The term 'traffic', which is used to describe the flow of telephone calls, is described in more detail in Chapter 6.) The bottom of the hierarchy is the exchange serving its catchment area of telephone lines. These exchanges are known as 'local exchanges' in the United Kingdom (and 'Class 5 central offices' in the United States [4]). Similarly, telephone lines are known as 'local lines' or 'subscriber lines', and the collection of local lines as the 'local network'. At the second level of the hierarchy are the trunk exchanges, and at the top of the hierarchy is the trunk tandem exchange. This parenting of trunk exchanges on higher-level exchanges can continue further. However, in practice, telephone networks generally have no more than three levels of trunk exchanges: primary, secondary and tertiary. Note that only the local exchanges have subscriber lines attached; the various levels of trunk exchanges switch only between trunk and junction routes to and from other exchanges. The significance of this difference between local and all other types of exchanges will be considered further in Chapter 6. As an example of a typical public switched telephone network (PSTN), we will consider the simplified representation in Fig. 1.9. This shows the PSTN for BT (British Telecommunications), which covers all of the United Kingdom (except for the town of Hull). The structure is shown as a hierarchy, with the exchanges serving subscriber lines forming the lowest level., there are three ways of connecting subscriber's lines to the network:

- Type (i) where subscribers are connected directly to the central local exchange (known as a 'processor node');

- Type (ii) where subscribers are connected to a remote unit (known as a 'remote concentrator'), which in turn is parented onto and controlled by the central local exchange;

- Type (iii) where subscribers are connected to an autonomous very small rural exchange, which in turn is connected to the central local exchange.

In BT's PSTN there are about 800 of the central local exchanges ('processor nodes'), about 5,200 remote units ('remote concentrators'), and some 480 very small rural exchanges. In total, these exchanges and units serve about 29million subscribers' lines. The (central) local exchanges are parented onto their trunk exchange (known as a 'digital main switching unit – DMSU'). This hierarchy has only one level of trunk switching, with all trunk exchanges having direct routes to all others. This arrangement ensures that any subscriber can be connected to any other on this network and that there will be a maximum routing of two trunk exchanges between the originating and terminating local exchanges. Calls between local exchanges, which are in the same locality and where there is sufficient demand, are carried over direct links – known as 'junction routes', as shown in Fig. 1.9. The number of junction routes is minimised in large urban areas, such as London, Birmingham and Manchester by the use of junction tandem exchanges (known as 'digital junction switching unit – DJSU'). Further network optimization is achieved by offloading the trunk exchanges of calls flowing just within the region by switching these at regional tandems (known as 'wide area tandem – WAT'). BT's PSTN has some 15 junction tandems (JTs), about 20 regional tandems and around 80 trunk exchanges. In addition to providing full interconnectivity between all subscribers on its network, BT's PSTN also provides access to international exchanges within the United Kingdom for calls to other countries. As Fig. 1.9 shows the international exchange is above the trunk exchange in the hierarchy. BT is one of several network operators in the United Kingdom that have international exchanges (known as 'international switching centre – ISC'). The routing of international calls is considered in more detail in Chapter 2.

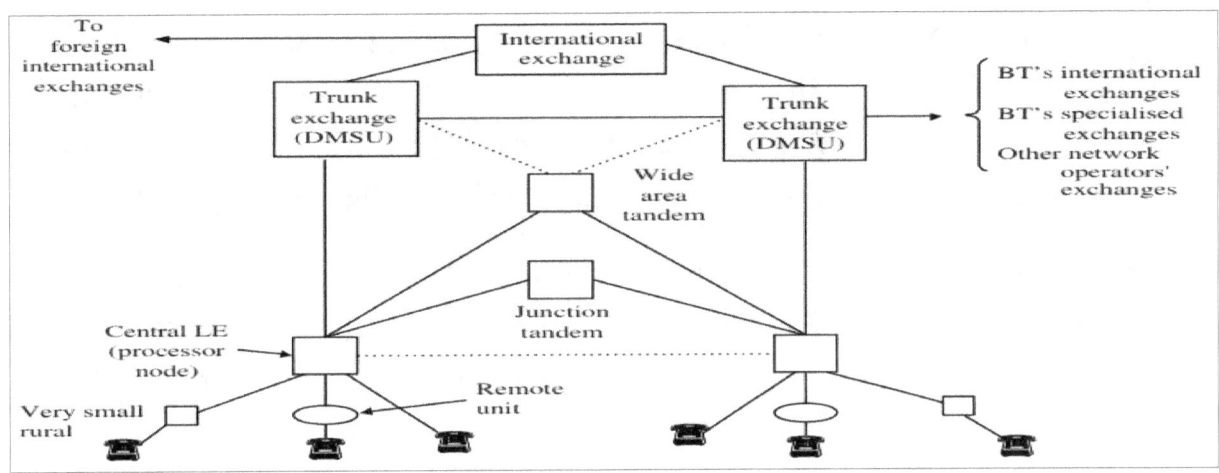

Figure 1.9 BT's Public Switched Telephone Network

The PSTN of BT also acts as a collector of calls from its subscribers to the networks of other operators (fixed and mobile) in the United Kingdom, and similarly as a distributor of calls from other operator's networks to subscribers on the BT network (Fig. 1.9). The points of interconnection (POI) between BT's PSTN and the other networks are at the trunk exchanges. Similarly, the trunk exchanges also act as collectors and distributors of traffic between PSTN subscribers and the variety of specialized BT exchanges covering functions such as operator services and directory enquiries, Centrex and VPN service for business customers, and network intelligence centers or the variety of calls that need more complex control than the PSTN exchanges can provide.

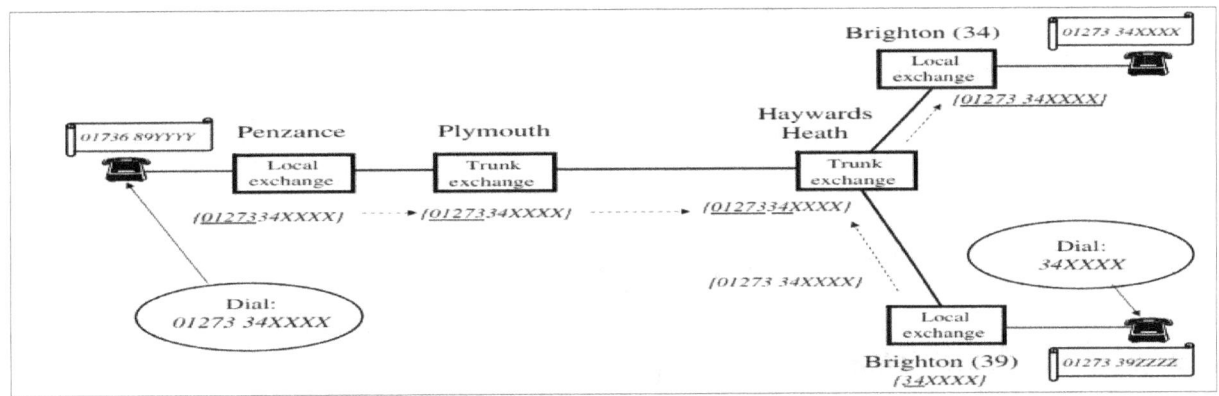

Figure 1.10 An Example of a Call Routing

SS7 Signaling

- Common Signaling System 7, also called SS7 or C7, was developed by the in order to increase the efficiency of the public voice system. SS7 is a separate network whose duties are setting up, tearing down, monitoring, and routing calls on the PSTN.

SS7 is akin to TCP/IP in that it operates at several layers of the OSI model. And, like TCP/IP, SS7 is packet-based. It is a software-based system that operates independently of the voice transport itself (the PSTN).

- SS7 works behind the scenes, so interacting with SS7 is something that the CO switch, not your phone or PBX, must do. SS7 is called an out-of-band signaling standard because, unlike DTMF, it doesn't use the same frequency band, or even the same transport, as the voice transmission.

- Out-of-band signaling is also called CCS, or common channel signaling. It's the technique used by all telecommunication vendors—including cellular phone service providers, long-distance companies, and local exchange carriers (LECs). All of these networks share one thing in common: a common bond in SS7.

SS7 and PSTN

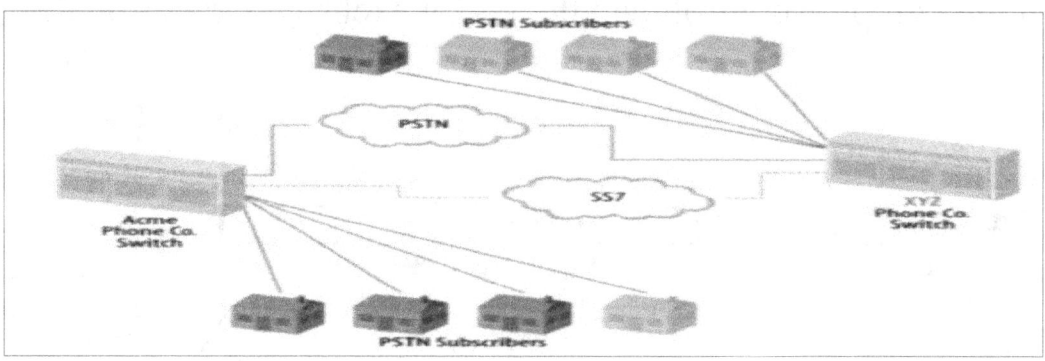

Subscriber Loop Design

- Any use of telephone channels involves two unidirectional paths, one for transmission and one for reception.

- The local loop, which connects a telephone to a local exchange is a two-wire (2W) circuit that carries the signal in both transmission directions.

- Even asymmetrical digital subscriber lines (ADSLs) use this same 2W local

- To connect a 2W local loop to a 4W network a circuit called a 2W/4W hybrid is needed.

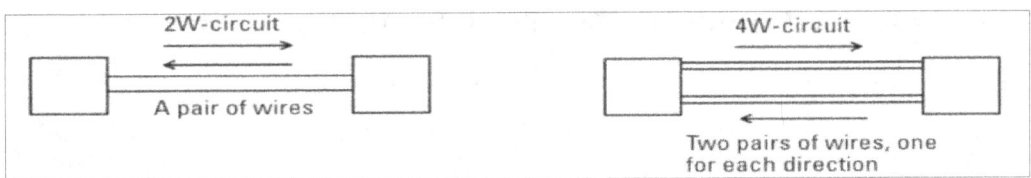

Normal Signal Flow

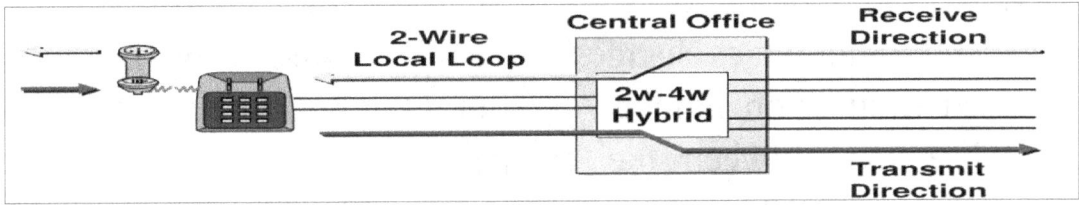

- 2- to 4-wire hybrid combines receive and transmit signals over the same pair

- 2-wire impedance must match 4-wire impedance

1.4 How does a network set up a call connection?

A fundamental requirement of routing through any telecommunications network, whether it is to carry a voice call or a packet of data, etc. – is that each termination (or end point) on the network has an address, which can be used to indicate the desired destination. In the case of telephony each subscriber's line has a permanent unique number, which is used by the exchanges in the network as its address. We are now in a position to consider a call set up across the PSTN. Fig. 1.10 illustrates the routing of a national call within the United Kingdom from a subscriber in Penzance in the South West tip to a subscriber in Brighton on the South coast. As an example, we assume that the subscriber in Penzance, 01736 89YYYY (i.e. area code 01736, exchange code 89 and subscriber number YYYY), dials their friend in Brighton, 01273 34XXXX (i.e. area code 01273, exchange code 34 and subscriber number XXXX). There are several local exchanges in the Brighton area, two of which – 34 and 39 – are shown in Fig. 1.10. On receiving the dialed digits from the caller, the control system of the local exchange in Penzance (the 'originating exchange') examines the first five digits, as shown by the underlining of the numbers in Fig. 1.10. The initial '0' indicates to the control system that a national number has been dialed. Since the control system does not recognize 1273 as either its own code or that of any other exchanges to which it has direct routes, the call is switched through to its parent trunk exchange in Plymouth. This is achieved by the sending of signals between the control systems in Penzance and Plymouth exchanges (shown by dotted arrows), conveying the required destination number. The number information sent in the forward direction between the exchanges is shown between curly brackets in Fig. 1.10. The control system in Plymouth exchange examines digits 01273 and concludes that this is not one of its dependent area codes, but is owned by Haywards Heath trunk exchange. In the UK network all BT trunk exchanges are fully interconnected and so the control system of Plymouth exchange selects a direct route to Haywards Heath and signals the full destination number. Haywards Heath exchange control system recognizes 01273 as one of its dependent area codes and so it examines the next two digits, 34, to determine the destination local exchange. The Haywards Heath trunk exchange then routes the call to the Brighton-34 exchange. At the destination exchange, the control system on recognizing 0123 34 then examines the final XXXX to determine the called subscriber number. In the case of a local call from the other Brighton exchange, 39, the set-up sequence is simpler, as shown in Fig. 1.10. The calling subscriber dials only the local number since the destination number has the same area code. The control system of Brighton-39 on examining the first two digits, 34, recognizes that the call is destined for one of the other exchanges dependent on the Haywards Heath trunk exchange, and so routes the call accordingly. Notice that, in this case of a local call, the control system inserts the appropriate area code and signals the full destination national number to its trunk exchange. Haywards Heath trunk exchange then uses the standard decoding procedure to determine that the call is destined for Brighton-34 exchange; the call is then completed as before. The dialed number is not only used by the exchanges to route the call through to its destination, but importantly the number is also used to determine the appropriate charge rate for the call. In both examples above it is the parent trunk exchange that determines the charge rate for the call and indicates this to the originating local exchange, which then times the call and records the incurred charge in its control system memory. This is later downloaded to produce an entry on the calling subscriber's telephone bill.

Telephone Numbering:

- The numbering is hierarchical, and it has an internationally standardized country code at the highest level.

- An international prefix or international access number is used for international calls. It tells the network that the connection is to be routed via an international telephone exchange to another country.

- The country code contains one to four numbers that define the country of subscriber B. Country codes are not needed for national calls because their purpose is to make the subscriber identification unique in the world. A telephone number that includes the country code is called an international number and it has a maximum length of 12 digits

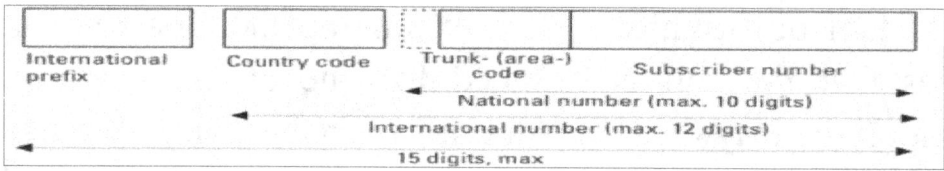

1.5 Waveforms

It is important at this point to ensure an understanding of the basic concept of waveforms and how they will be used throughout the remainder of the book. In general, all communication systems are characterized by the range of frequencies that can be carried, usually termed the 'bandwidth' of the system. Fig. 1.11 shows the classical sinusoidal waveform showing the regular oscillating pattern that occurs often in nature, e.g. a swinging pendulum. The time between successive peaks or troughs in the amplitude of the waveform (which might be length, voltage, power, etc.) is known as the 'period'. Each period comprises one full cycle of the waveform.

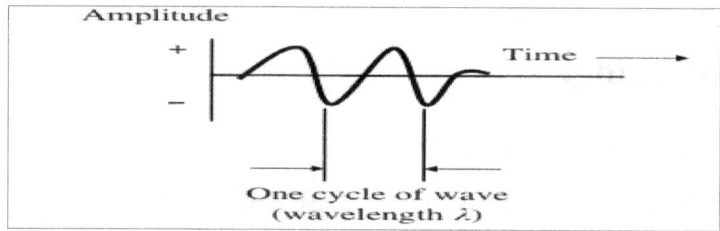

Figure 1.11 Sinusoidal Waveform

The distance travelled in this period of time is called the wavelength, given the symbol λ. The number of periods per unit time is known as the frequency of the waveform, measured in cycles per second (cps), or to use the international standard, Hertz (Hz) [6]. The sinusoidal waveform of Fig. 1.11 in the case of a sound wave would represent a single tone. For example, a single sinusoidal sound wave of 256 Hz produces the well-known musical note of middle C. However, speech and music are a mixture of many different tones, each with different frequencies and amplitudes, resulting in complex waveform shapes, as shown in Fig. 1.1(a). Within this book two fundamentally different forms of waves are considered. The first is that of sound waves, which are moving physical vibrations in a substance. These can be heard and felt by humans. Thus, sound waves are carried through the air in the form of vibrating air molecules, e.g. from a person speaking to a listener, or a guitar string being plucked. Sound waves can also be carried as vibrations through a wooden door, through water or as vibrations along a railing which someone at the far end is banging, or as vibrations along the railway lines as the distant train approaches. Sound is generated by vibrating the air, as is apparent when a finger touches a loud speaker of a HiFi system at full volume! Vibrations carried along the surface of the sea (i.e. 'sea waves') are another example of sound wave. The other form of waves are electromagnetic waves. Light is the most obvious example of electromagnetic radiation. Light travels as very high-speed waves of electric and magnetic forces. Other examples of this type of waveform are radio waves (carrying radio and TV broadcasts, etc.), microwaves and X-Rays. They are all examples of the phenomenon of electromagnetic radiation, but at different frequencies. Thus, within telecommunications networks, it is electromagnetic waveforms that are carried as electricity over metallic wires, or as radio waves through the air from radio masts to a mobile handset, or as light waves through optical fiber cables. The full range of the electromagnetic radiation spectrum is illustrated in Fig. 1.12. The key point is that the sound waves emitting from a speaking person are converted into electromagnetic waves for the conveyance through the

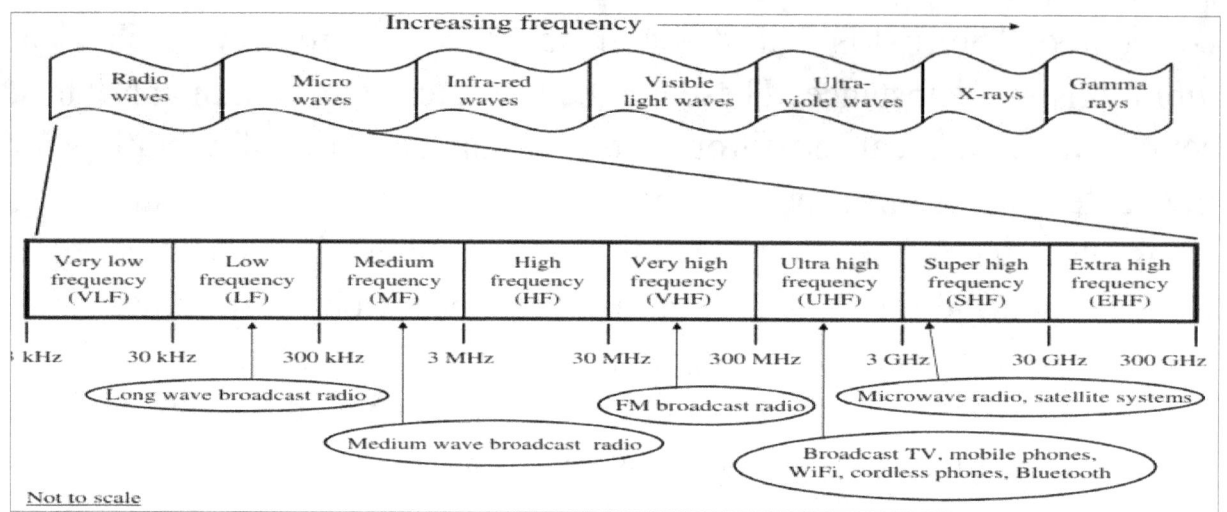

Figure 1.12 The Electromagnetic Radiation Spectrum

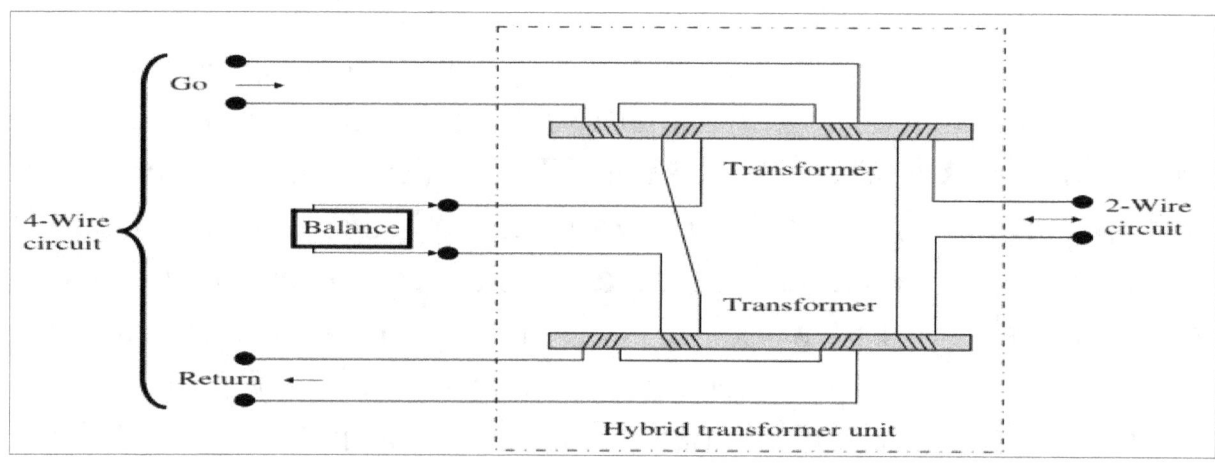

Figure 1.13 Hybrid Transformer

telecommunications network. (Thus, in Figs 1.1 and 1.2, the sound waves are carried over the pair of wires as an electromagnetic signal over the wires – and converted back to sound waves at the far end.) The sound or speech waveform (occupying the frequency range of 0–4 kHz as described in Chapter 3) can be carried by an electromagnetic wave at any appropriate frequency level. Thus, if carried straight over a copper pair the sound waveform is converted to an electrical analogue signal occupying the same frequencies (0–4 kHz); or they may be carried at higher frequencies within multiplex signal (as described in Chapter 3), or at radio frequencies if carried over long wave, medium wave or FM radio, or at even higher frequencies in the electromagnetic spectrum if carried as part of a TV broadcast, and so on up to the highest frequencies when the speech is carried as light waves over an optical fiber cable.

Transmission Systems

- Link characteristics
  - information carrying capacity (bandwidth)
    - information sent as symbols
    - 1 symbol >= 1 bit
  - propagation delay
    - time for electromagnetic signal to reach another end
    - light travels at 0.7c in fiber ~8 microseconds/mile
    - NY to SF => 20 ms; NY to London => 27 ms
  - attenuation
    - degradation in signal quality with distance
    - long lines need regenerators
    - optical amplifiers are here

Transmission Impairments

- Signal Attenuation
- Interference
- Coupling between wires
- Near-end crosstalk (NEXT) (From TX to RX at a common location)
- Far-end crosstalk (FEXT) (From TX to RX at a distant location)
- Noise

- Thermal noise - White noise with a Gaussian (Normal) distribution of amplitudes
- Noise measurement is important
- Standard reference value is 1 pW → -90 dBm

Power Levels

- Read the dB Tutorial on the course web site
- The delivered signal power must be high enough to be clearly perceived. Not so strong that echo and singing result
- Transmission links are designed with specific amount of net loss. Via net loss (VNL)
- Transmission Levels Point (TLP) are used as a convenient means of expressing signal loss or gain within a circuit.
  - The TLP is a point in the circuit expressed as the ratio (in dB) of the power of the signal at that point to the power of the signal at a reference point (0 TLP).
  - TLP is the measurement of the signal gain or loss relative to the 0 TLP.
  - dBm0 = Signal Power (dBm) - TLP (dB)
  - "0" indicates that the specification is relative to the 0-TLP.
  - Ex: If an absolute noise power of 100 pW (20 dBrn or -70 dBm) is measured at a -6 TLP, it is expressed as 26 dBrn0.

dB Applied to the Voice Channel

- Noise and amplitude distortion
- Amplitude distortion is the same as frequency response.
- The noise annoys the listener. How much noise will annoy the average listener?

- The human ear is a filter as is the telephone earpiece
- Amount of annoyance of the noise to the average listener varies
- We "shape" the VF channel as a function of frequency
- Weighting curve
- C-message response (NA)

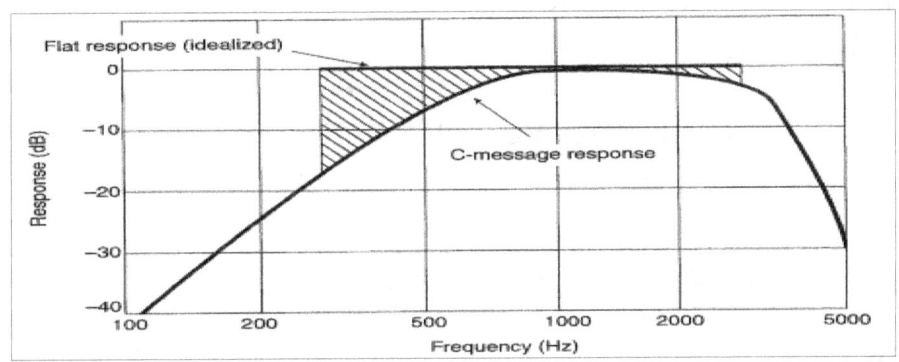

dB Applied to the Voice Channel

- The lowest discernible signal that can be heard by a human being is -90 dBm (800 or 1000 Hz)
- If noise power is measured with C-message weighting, dBrnC0 is used.
- 0 dBrnC=-92 dBm (with white noise loading of entire voice channel)

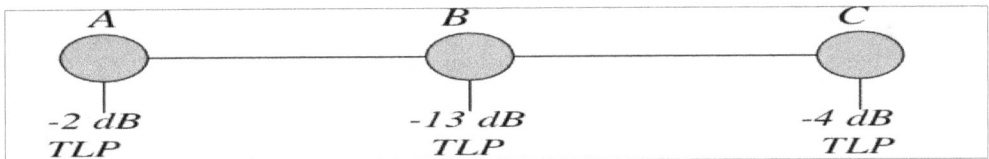

Example:

Using the above figure, determine each of the following: (a) the signal power to be applied at point B to determine if points A and C are at the proper levels; (b) the amount of gain (loss) a signal experiences when propagating from A to C; and (c) the amount of noise that would be measured at C if 27 dBrnC of absolute noise is measured at B and no additional noise occurs on the B-to-C link.

Solution: (a) Because point B is -13 dB TLP, the proper test tone level is -13 dBm (0.05 mW) (b) Because the TLP values drop by 2 dBm, there is 2dB net loss from A to C. (c) An absolute measurement of 27 dBrnC at B is 40 dBrnC0. This is also 40 dBrnC0 at C. The absolute noise power measured at C would be 40-4=36 dBrnC.

1.6 Other forms of telephone networks

Concepts

- Single basic service: two-way voice
  - low end-to-end delay
  - guarantee that an accepted call will run to completion

1. Endpoints connected by a circuit
  - like an electrical circuit
  - signals flow both ways (full duplex)
  - associated with bandwidth and buffer resources

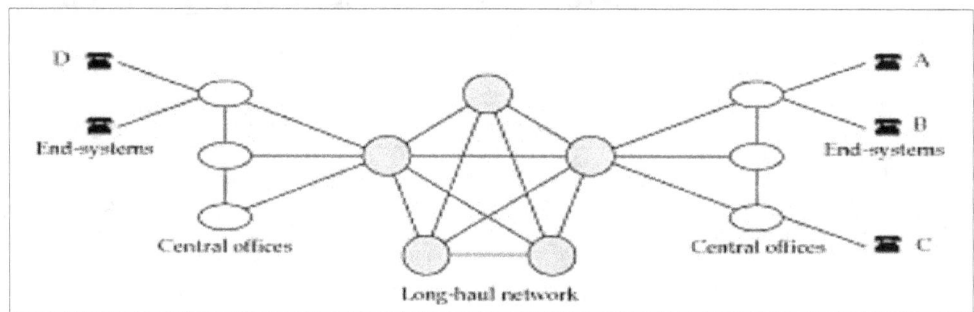

- Fully connected core
  - simple routing
  - telephone number is a hint about how to route a call but not for 800/888/700/900 numbers

> hierarchically allocated telephone number space

The pieces

1- End systems

> Transducers: key to carrying voice on wires

> Dialer

> Ringer

> Switch hook

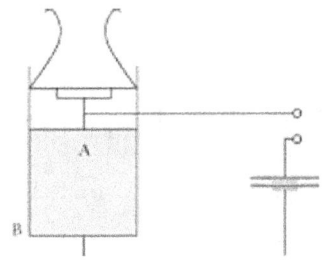 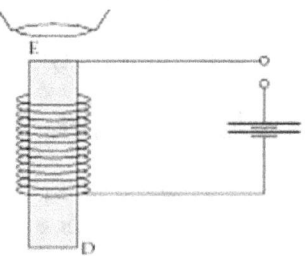

2. Transmission:

- **Transmission: Multiplexing**

    > Trunks between central offices carry hundreds of conversations

    > Can't run thick bundles!

    > Instead, send many calls on the same wire, multiplexing

    > Analog multiplexing (Frequency Division Multiplexing), bandlimit call to 4 KHz and frequency shift onto higher bandwidth trunk. Obsolete

- Digital multiplexing

    > first convert voice to samples

    > 1 sample = 8 bits of voice

    > 8000 samples/sec => call = 64 Kbps

- **Time division multiplexing**
  - trunk carries bits at a faster bit rate than inputs
  - n input streams, each with a 1-byte buffer
  - output interleaves samples
  - need to serve all inputs in the time it takes one sample to arrive
  - => output runs n times faster than input
  - overhead bits mark end of frame
  - Multiplexed trunks can be multiplexed further
  - Need a standard
  - US/Japan standard is called Digital Signaling hierarchy (DS)

Digital Signal Number	Number of previous level circuits	Number of voice circuits	Bandwidth
DS0		1	64 Kbps
DS1	24	24	1.544 Mbps
DS2	4	96	6.312 Mbps
DS3	7	672	44.736 Mbps

- **Transmission: Link technologies**
  - Many in use today
  - twisted pair
  - coax cable
  - terrestrial microwave
  - satellite microwave
  - optical fiber
  - Increasing amount of bandwidth and cost per foot
  - Popular
  - fiber

- satellite

- **Transmission: fiber optic links**
  - Wonderful stuff! lots of capacity, nearly error free, very little attenuation, hard to tap
  - A long thin strand of very pure glass

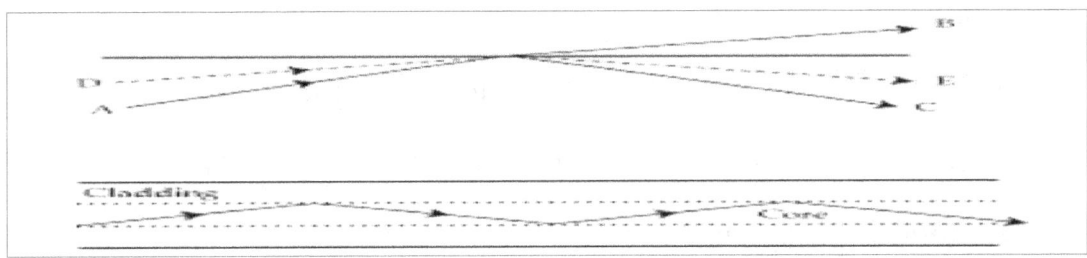

- **Transmission: satellites**
  - Long distances at high bandwidth, Geosynchronous, 36,000 km in the sky, up-down propagation delay of 250 ms, bad for interactive communication, slots in space limited, Non-geosynchronous (Low Earth Orbit), appear to move in the sky, need more of them, handoff is complicated, e.g. Iridium.

3. **Switching**

- Problem:
  - each user can potentially call any other user
  - can't have direct lines!
  - Switches establish temporary circuits
  - Switching systems come in two parts: switch and switch controller

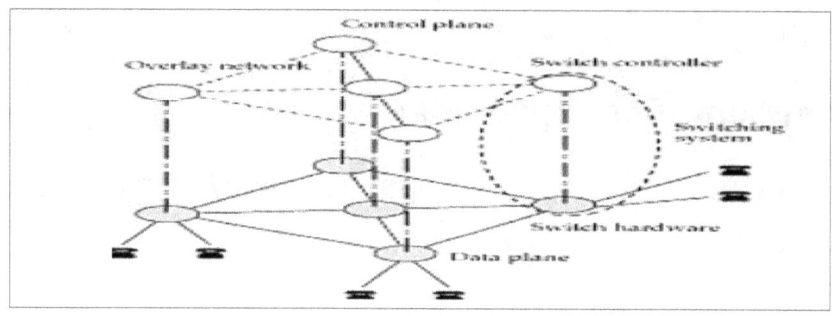

- what does a switch do?
  - Transfers data from an input to an output
  - many ports (up to 200,000 simultaneous calls)`
  - need high speeds
  - Some ways to switch: space division.
  - if inputs are multiplexed, need a schedule (why?) Signaling

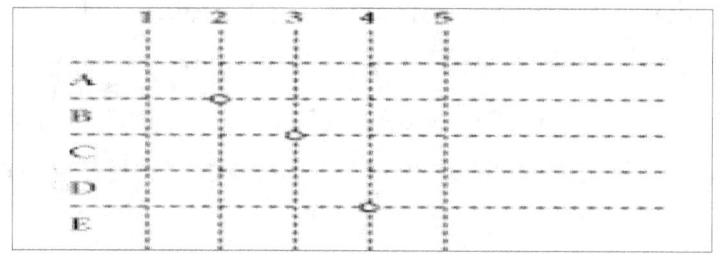

Another way to switch

- time division (time slot interchange or TSI), also needs scheduling
- To build larger switches we combine space and time division switching elements

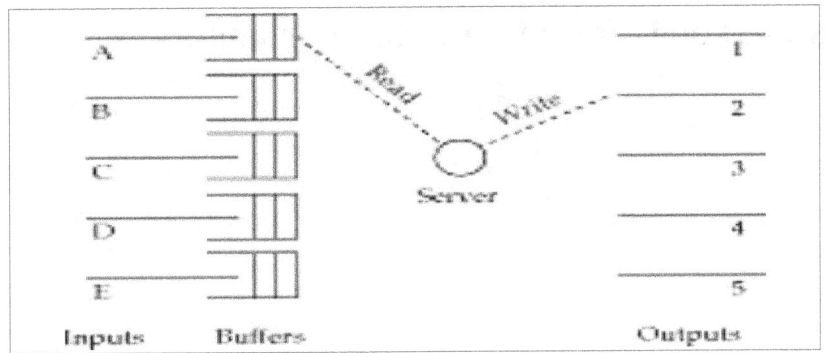

4. Signaling

- Recall that a switching system has a switch and a switch controller
- Switch controller is in the control plane
  - does not touch voice samples
- Most common control signals
- Dial tone, ringback, and busy tone
- Supervisory (conveying status) & information bearing signals
- Manages the network
  - call routing (collect dialstring and forward call)
  - alarms (ring bell at receiver)
  - billing
  - directory lookup (for 800/888 calls)
- Signaling network
  - Switch controllers are special purpose computers
  - Linked by their own internal computer network. Common Channel Interoffice Signaling (CCIS) network
  - Earlier design used in-band tones, but was severely hacked
  - Also was very inflexible

> Messages on CCIS conform to Signaling System 7 (SS7) spec.

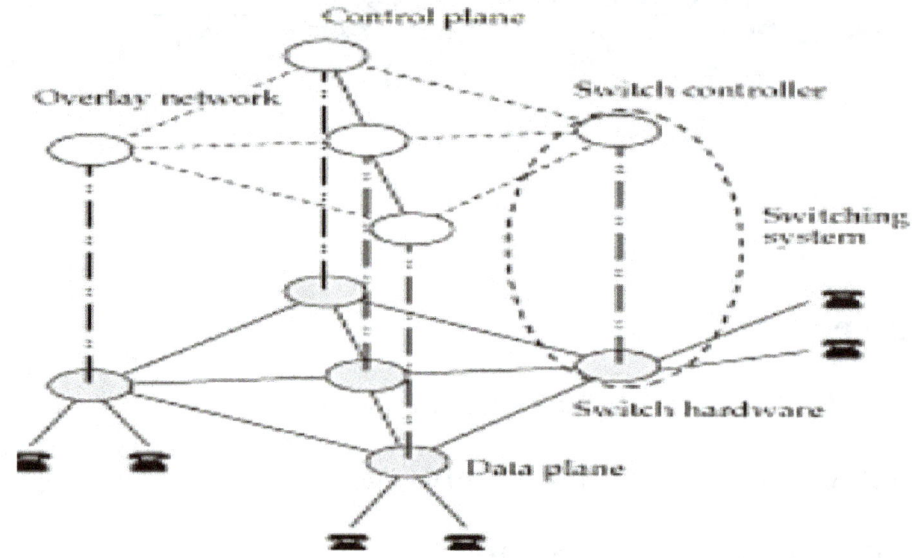

Cellular communication:

- Mobile phone talks to a base station on a particular radio frequency
- Aren't enough frequencies to give each mobile a permanent frequency (like a wire)
- Reuse
- temporal
- if mobile is off, no frequency assigned to it
- spatial, mobiles in non-adjacent cells can use the same frequency

Mobile networks: The concept of a cellular mobile telephone network is shown in Fig. 2.1. The principle of operation is similar to that of the PSTN, although there are some important differences. Obviously, is

- that a two-way radio link is used to connect the mobile handset to its exchange, rather than a copper pair of wires, so that the user has freedom of movement.

This ability of the handset to move around means that the serving exchange must have a system to identify a calling handset and another system to keep track of where a particular handset is at any time, so that it can send or receive calls.

The basis of the 'access network' is a set of cells. These cells form an area ranging from about 1–10 km radius and have a centrally located radio transmitter/receiver (also known as a 'transceiver'), collocated with a base station (BS). A two-way radio link is potentially available to all handsets in the cell area. Groups of radio-channel pairs (Go and Return), each provided by a separate set of frequencies, are pre-assigned to the cell, and a pair of radio channels is allocated on demand by the base station controller (BSC) to a mobile handset wishing to make or receive a call. A BSC serves a catchment area of several cells with their associated BSs. The BSs are connected to their BSC by a fixed transmission link carried over either a point-to-point microwave radio system or an optical fibre cable. During the call, the handset is free to move within the cell. However, if the handset travels towards the boundary of its cell during a call the weakening signal from the BS and the strengthening radio signal from the adjacent cell's BS is detected by the BSC and a 'handover' between the two is managed without interruption to the call. This requires changing the send and received channels to those of the new cell. (If a spare set of radio channels is not currently available within the new cell, the call will have to drop out.) At the end of the call, or if the terminal moves into another cell's area, these radio channels are available for use by other calls. The mobile switching center (MSC), which is very similar to a large local PSTN exchange (but without the subscriber terminations), performs the switching of the mobile calls between all handsets operating within its catchment area of BSs or to other MSCs on its network. The MSC is also associated with control systems providing terminal-location management and authentication.

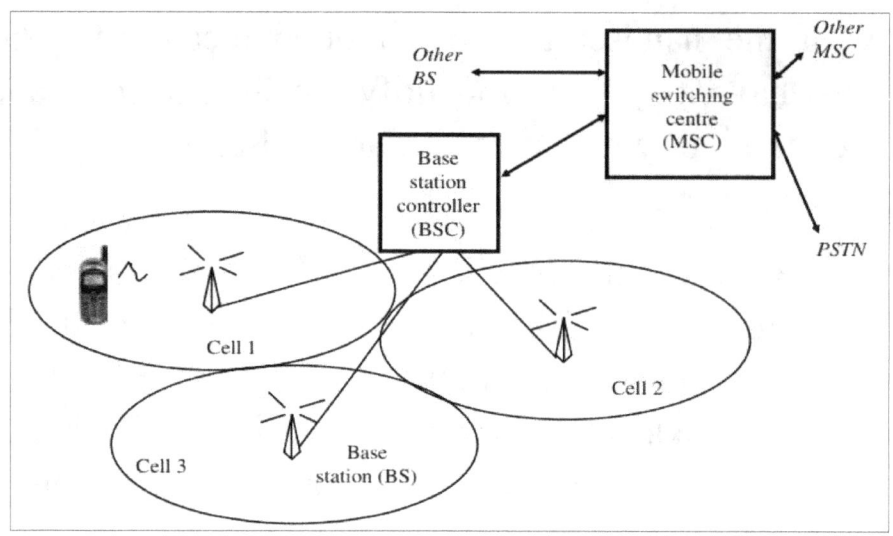

Mobile Cellular Network Concept

- In cellular technology, the region is divided into smaller cells. In each cell, a cell site serves cellphones in the cell
- Channels can be reused in different cells. Channel reuse supports more customers. This is the reason for using cells. (Channel 47 is reused in cells A, D, and F)

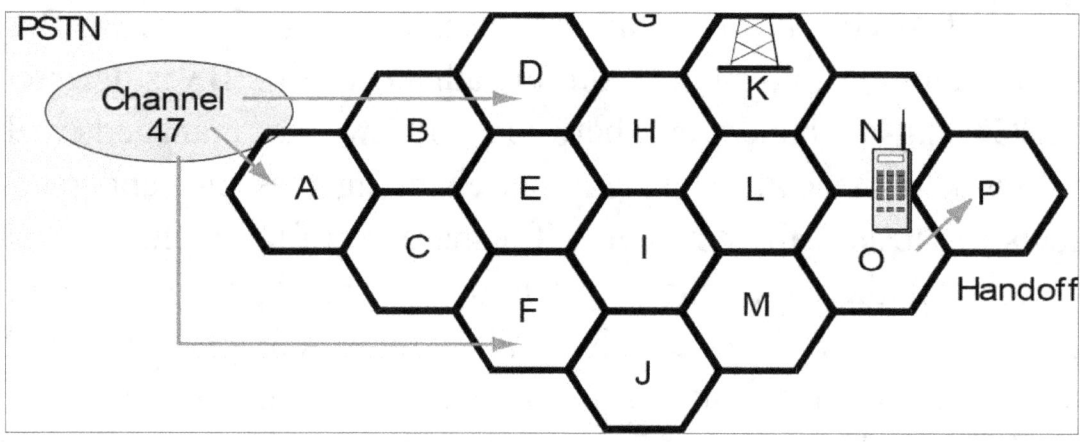

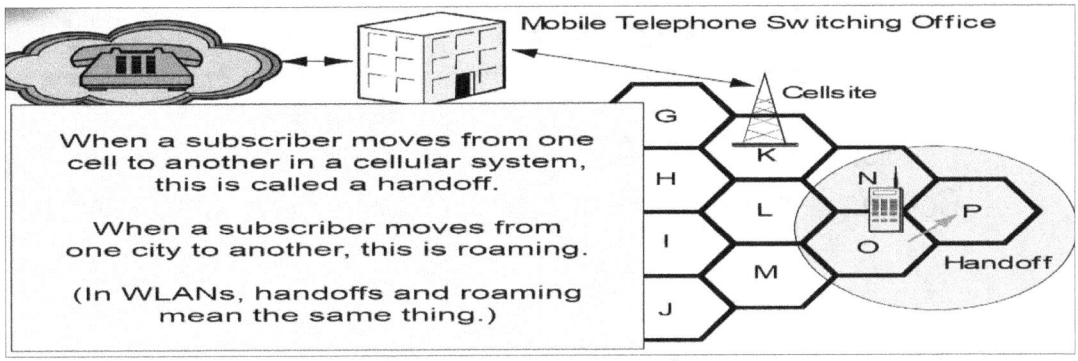

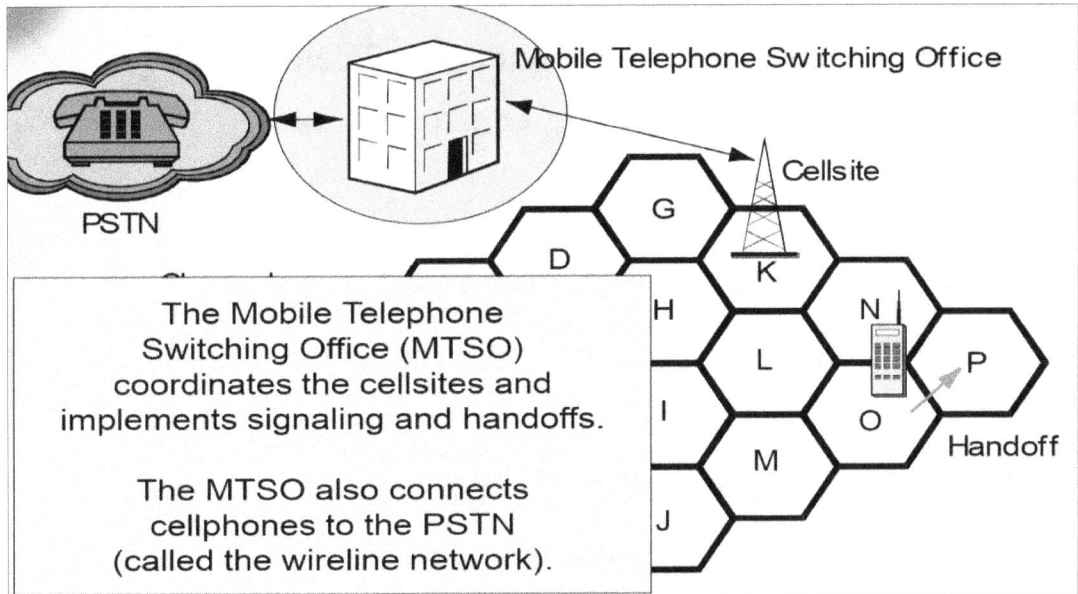

GSM is the worldwide standard for cellular voice:

- Uses time division multiplexing (TDM)
- Uses 200 kHz channels
- Divides each second into many frame periods
- Divides each frame into 8 slots
- Gives same slot in each frame to a conversation

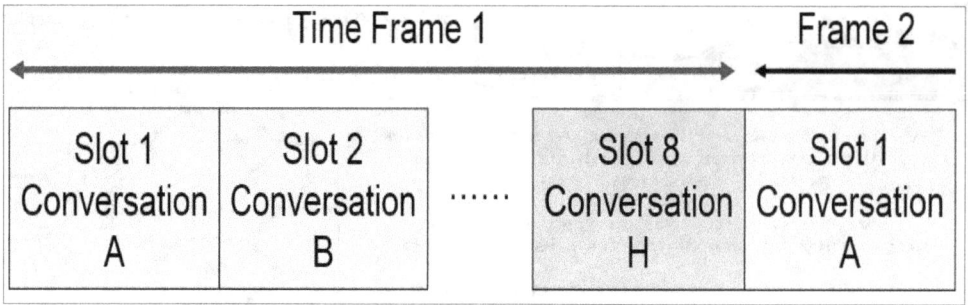

Cannot use the same channel in adjacent cells

- So can only reuse a channel about every 7 cells
- For example, suppose there are 50 cells
- Channel can be reused 50 / 7 times
- This is 7 (not precise, so round things off)
- So, each channel can support 7 simultaneous customers in these 7 cells

Code Division Multiple Access (CDMA)

- Also used in the United States
- A form of spread spectrum transmission
- Unlike traditional spread spectrum technology, multiple users can transmit simultaneously
- 1.25 MHz channels
- Can support many users per channel
- Can use the same channel in adjacent cells
- So can only reuse a channel in every cell

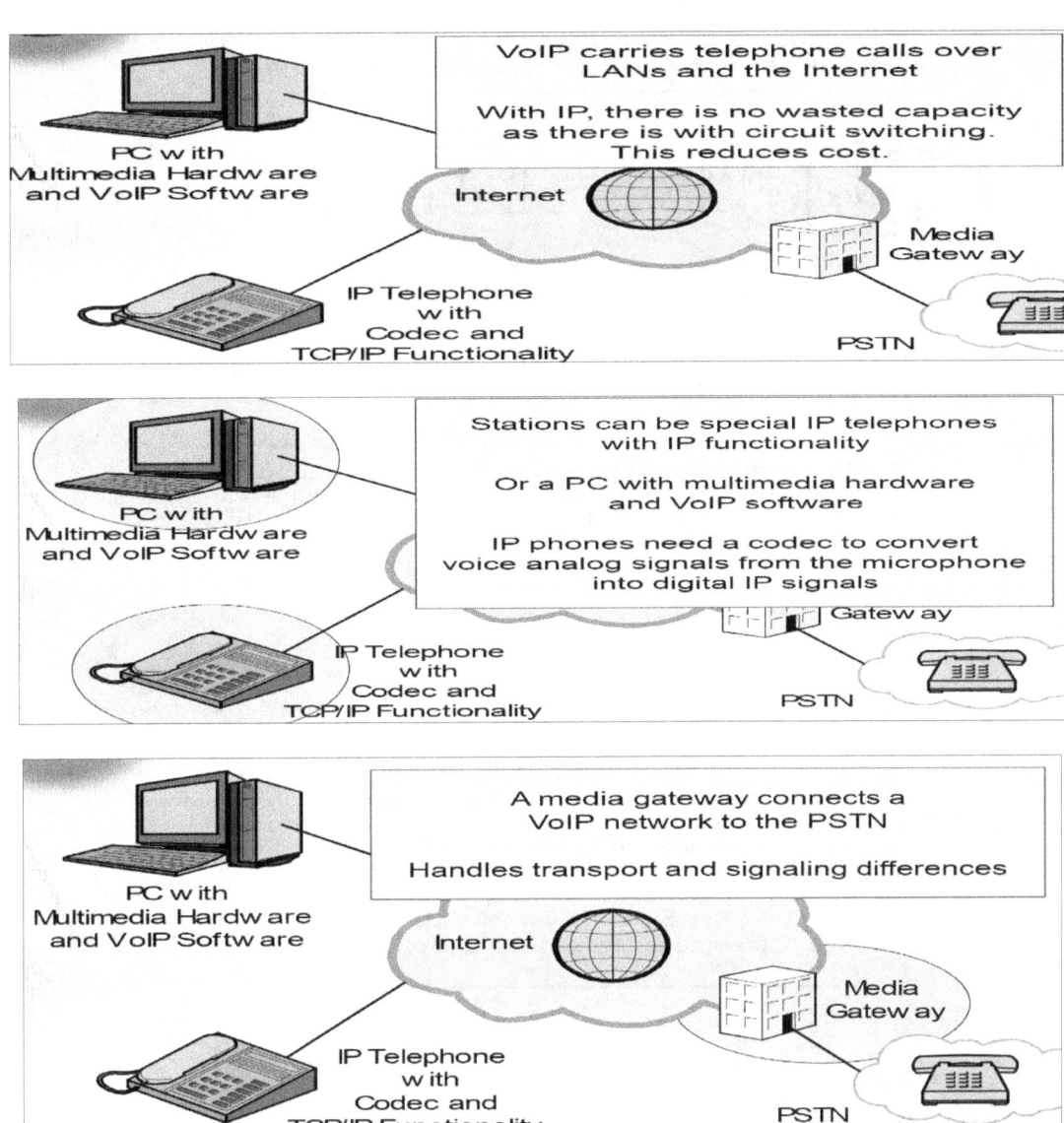

Speech Codes

Codec	Transmission Rate
G.711	64 kbps (pulse code modulation)
G.721	32 kbps (adaptive PCM)
G.722	46, 56, or 64 kbps
G.722.1	24, 32 kbps
G.723.1A	5.3, 6.3 kbps

There are several codec standards. They differ in transmission rate, sound quality, and latency. Both sides must use the same codec standard.

VoIP Protocols

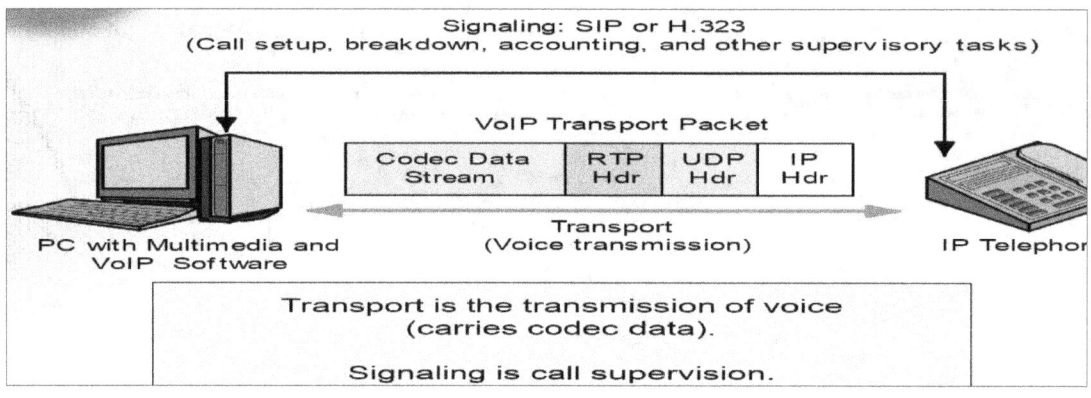

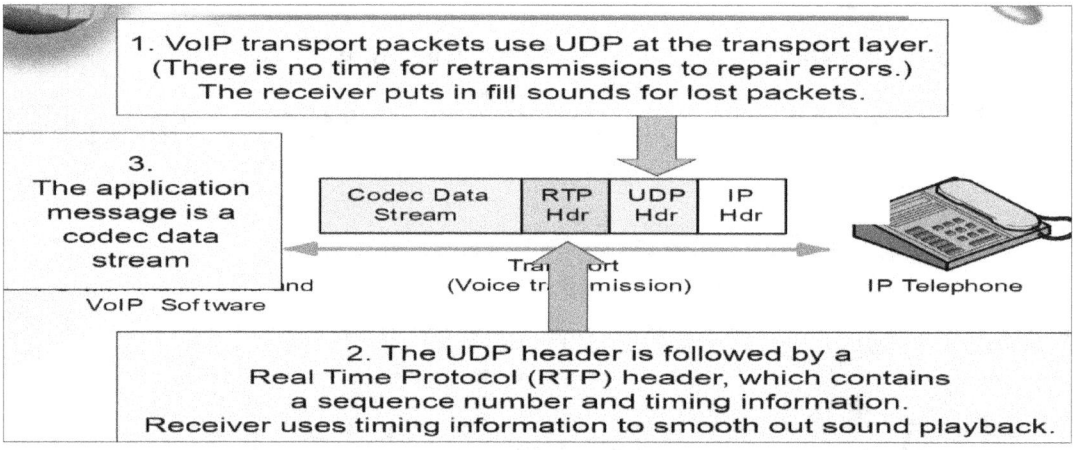

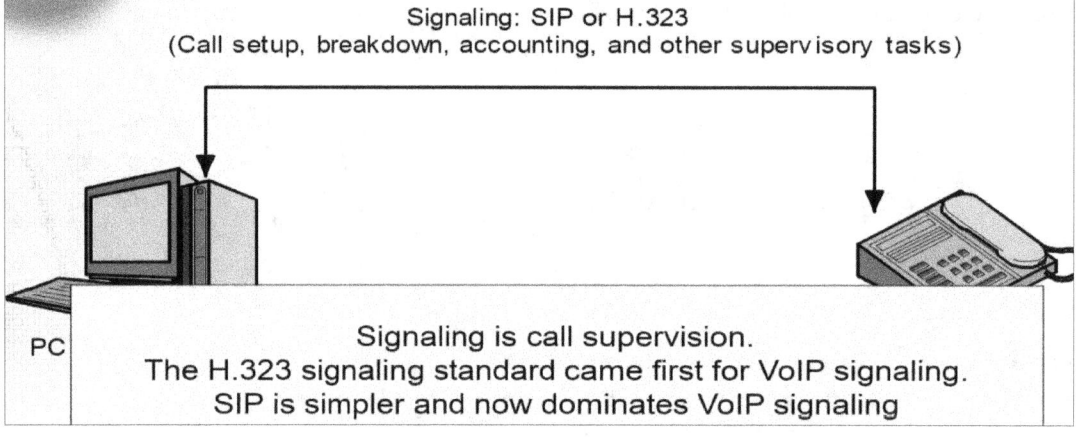

The Other VoIP

- It's not just voice over IP
- Video Telephones
- Video Conferencing
  - ➢ PC to PC

- > Multiparty
- > Sometimes room-to-room
- Video Downloads on Demand

Telegraph:

Definition: Telegraph (from Greek):

- tele- means "afar, far off", graphein- means "writing"
  - > Telegraph is an instrument that transmits the writings to a distant place

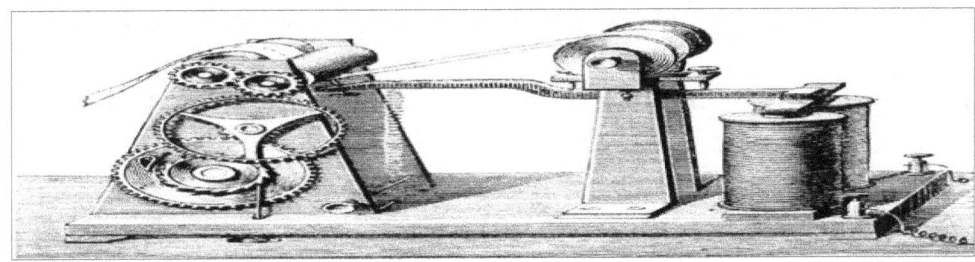

Before the invention of telegraph, there were the signaling systems that people could communicate over the long distance.

- The signaling systems were called "semaphore" which they used flags or lights. In this system, there was the observer who decoded a signal from a high tower and then sent it to the next station.

THE GROWTH OF AN IDEA

- In 1832, when Samuel F.B. Morse was in a ship returning from Europe, he accidentally had an idea of using electricity to communicate over distance. Because in this ship, many passengers discussed about the electromagnet which was recently invented. After Morse entirely understood how the electromagnet worked, he realized that sending a code-message via a wire might be possible.

- However, Morse found that this were difficult. Thus, he asked his friend, Leonard D. Gale in order that he could follow his idea.

- In 1837, because of the Panic which caused a long depression, Morse could not introduce his new system although he had enough confidence to do that.

- However, Morse used this time to travel to Europe not only to register a patent but also to examine the English telegraph systems. He felt very confident because even though his main competitor had created an ingenious mechanism, Morse's telegraph was more efficient and easier to use.

- Fortunately, in 1843, the economic was recovering, Morse again asked the Congress for the cash of $30,000 in order to build the underground telegraph line from Washington to Baltimore.

- Morse hired the great construction engineer, Ezra Cornell to lay the pipe which carried the wire. However, because of the wire was defective insulation, Cornell suggested stringing wires overhead on trees. Morse approved of this great idea.

- Finally, the line was completed.

Sending the first telegraph message

THE REMARKABLE EVENT
On May 24, 1844, the first message, "What hath God wrought?" was sent by the telegraph.

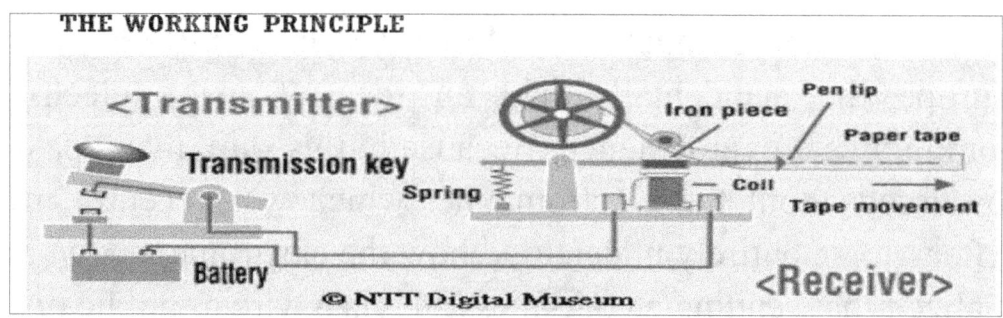

THE PROCESS OF DEVELOPMENT

- At first, the telegraph system progressed slowly and the system could not work for the entire country. Thus, Morse continually tried to improve the telegraph
- system as well as to extend the telegraph line. Fortunately, other companies began to open their own systems, the telegraph system began to develop.
- In 1914, automatic transmission was developed instead of trained code users that were used in the first telegraph to transmit the telegraph messages. This development in transmission made the message transmission much faster than the old way.
- At that time, all long-distance communication depended strongly on the telegraph.
- However, when the telephone and radio were invented, the telegraph was less developed than that in first haft of 1990s.

IMPACT

- With the invention of telegraph, the world suddenly became much smaller.
- With the invention of telegraph, the world became more connected.
- Moreover, the telegraph also contributed to the establishment of world peace by erasing the national–rivalries.

Lecture 2
Voice Transmission

2.1 Basic Concepts

In the simplest model of a telephone speech communication there is a direct, dedicated, physical connection between the two participants in the conversation, and this link is held for the duration of the conversation. The analogue electrical signal produced by the telephone at either end is sent on to connection without modification.

In Pulse Amplitude Modulation (PAM), the unmodified electrical signal is not sent on to the connection. Instead, short samples of the signal are taken at regular intervals, and these samples are sent on to the connection. The amplitude of each sample is identical to the signal voltage at the time when the sample was taken. Typically, 8,000 samples are taken per second, so that the interval between samples is 125μs, and the duration of each sample is approximately 4μs.

Because each sample is very short (~4μs) there is a lot of time between samples (~121μs). Samples from other conversations are put into this "spare time". Usually the samples from 32 separate conversations are put on to a single line. This process is called Time Division Multiplexing (TDM).

Each sample is very short, and will be distorted as it travels across a communications network. In order to reconstruct the original analogue, signal the only information the receiver needs to have about a sample is its amplitude, but if this is distorted then all information about the sample has been lost. To overcome this problem, the pulse is not transmitted directly, instead its amplitude is measured and converted into an 8-binary number - a sequence of 1s and 0s. At the receiver end, the receiver merely needs to detect if a 1 or a 0 has been received so that it can still recover the amplitude of a PAM pulse even if the 1s and 0s used to describe it have been distorted.

The process of converting the amplitude of each pulse into a stream of 1s and 0s is called Pulse Code Modulation (PCM)

Note that the process of PAM and PCM (but without the use of TDM) is essentially used to store music and speech on CDs, but with a higher sample rate, more bits per sample and complex error correction mechanisms.

Some terms are:

Sampling The process of measuring the amplitude of a continuous-time signal at discrete instants. It converts a continuous-time signal to a discrete-time signal.

Quantizing Representing the sampled values of the amplitude by a finite set of levels. It converts a continuous-amplitude sample to a discrete-amplitude sample.

Encoding Designating each quantized level by a (binary) code.

Sampling and quantizing operations transform an analogue signal to a digital signal.

Use of quantizing and encoding distinguishes PCM from analogue pulse modulation methods.

The quantizing and encoding operations are usually performed in the same circuit at the transmitter, which is called an Analogue to Digital Converter (ADC). At the receiver end the decoding operation converts the (8 bit) binary representation of the pulse back into an analogue voltage in a Digital to Analogue Converter (DAC)

Time Division Multiplexing (TDM) – Principle:

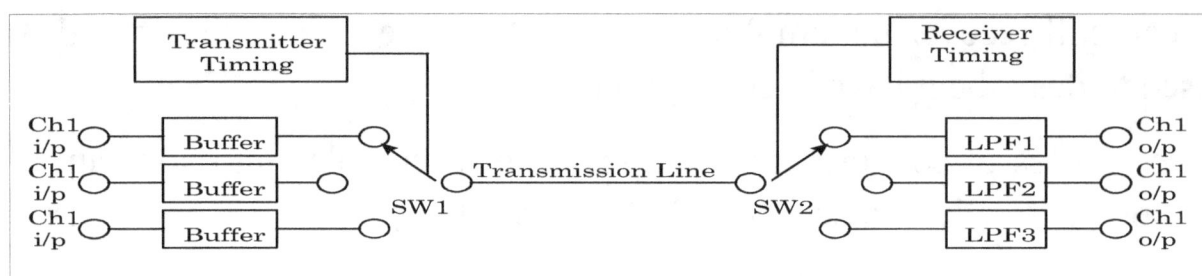

When sending samples of a signal instead of the signal itself there is time available between each of the samples. Samples from other analogue signals can be put into this space. The process of splitting up the time into slots and putting different signals into the time slots is known as Time Division Multiplexing (TDM). A basic real TDM system interleaves 32 signals and uses electronic switches. This is a diagram of a **3 channel PAM-TDM system.**

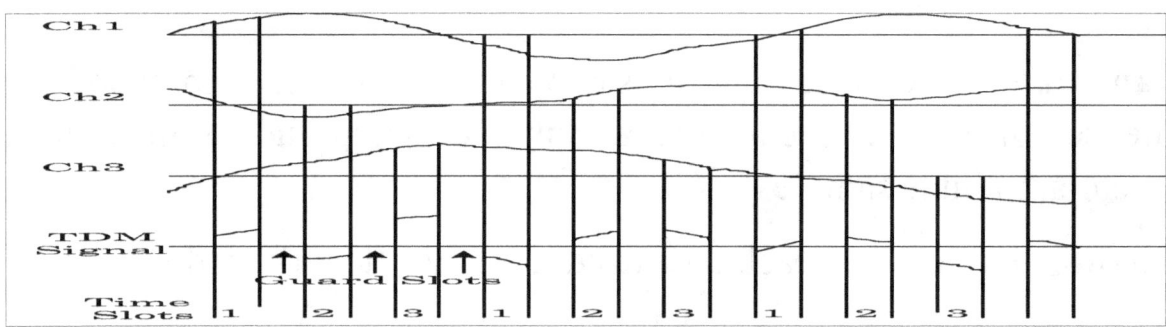

This diagram shows the waveforms produced during the operation of the PAM-TDM system. The switches connect the transmitter and the receiver to each of the channels in turn for a specific interval of time. In effect each channel is sampled and the sample is transmitted. When the switches are in the channel 1 position, channel 1 forms a PAM channel with an LPF for reconstruction, and so on for channels 2 and 3. The result is that the amplitudes samples from each channel share the line sequentially, becoming interleaved to form a complex PAM wave, as shown above. A major problem in any TDM system is the **synchronisation** of the transmitter and receiver timing circuits. The transmitter and receiver must switch at the same time and frequency. Also, SW1 must be in the channel 1 position when SW2 is in the channel 1 position, so that the switches must be synchronised in position also. In a system that uses analogue modulation (PAM) the time slots are separated by guard slots to prevent crosstalk between channels.

Sampling Theorem

Consider a band-limited signal with no frequency components above a certain frequency f_m. The sampling theorem states that this signal can be recovered completely from a set of samples of its amplitude, if the samples are taken at the rate of **$f_s > 2f_m$** samples per second.

This is often called the uniform sampling theorem for baseband or low-pass signals. The minimum sampling rate, $2f_m$ samples per second, is called the Nyquist sampling rate (or Nyquist frequency); its reciprocal $1/(2f_m)$ (measured in seconds) is called the Nyquist interval.

$f_s = 2 * f_m$ is called the Nyquist sampling rate.

For telephone speech the standard sampling rate is 8 kHz (or one sample every 125 μs).

Sampling Methods

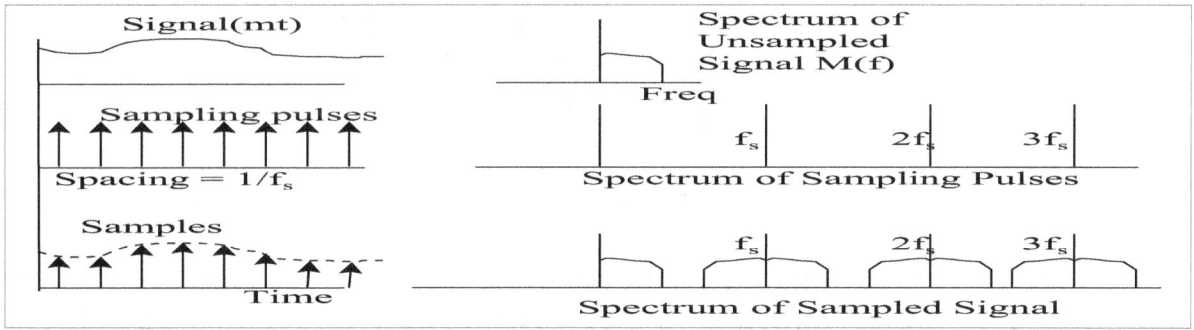

Suppose we have an arbitrary signal (the **baseband signal m(t)) which has a spectrum M(f).** Take infinitesimally short samples of the signal m(t) at a uniform rate once every t_s seconds i.e. at a frequency f_s. This is the ideal form of sampling; it is called instantaneous (or impulse) sampling. In effect the signal m(t) is multiplied by a train of impulses giving rise to a train of pulses as in the lower line of the diagram. The train of sampling impulses has a frequency spectrum consisting of all harmonics or multiples of f_s and all are at the same amplitude.

This **sampled signal has a spectrum as shown where M(f) is repeated unattenuated periodically and appears around all multiples of the sampling frequency ($f_s = 1/t_s$).**

To recover m(t) from the sampled signal we need only pass the sampled signal through a low pass filter with a stop frequency of $f_s/2$. All of the higher frequency components will be dropped. In the diagram, if f_s is greater than twice the highest frequency in m(t) the repetitions of the sampled spectra around the harmonics of the f_s do not overlap.

Flat - top Sampling

An Analogue to Digital Converter requires that the sample value be held constant for a fixed time until the conversion is completed. This requires a flat-top sampled signal. This has approximately the same repeated frequency spectrum as with the instantaneous sampling above, but with each repetition slightly spread out.

The simplest and most common sampling method is performed by a functional block termed a Sample and Hold (S/H) circuit.

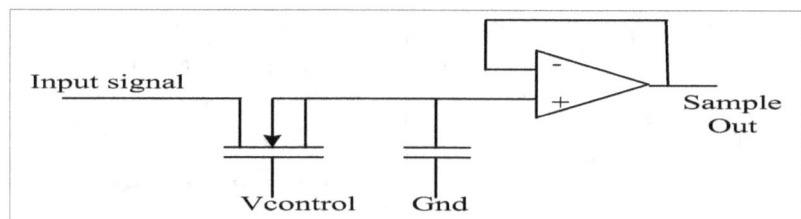

The output from the circuit must be held at a constant level for the sampling duration. Vcontrol switches the MOSFET ON until the charge on C is equal to the amplitude of the sampled voltage. Vcontrol then goes LOW, the MOSFET is OFF and the charge is held by the capacitor. The charge held on the capacitor puts a voltage across the capacitor, and it is held at that value until the next time that Vcontrol switches the MOSFET ON. This is called a sample and hold circuit and is usually used as the input to an ADC.

Aliasing Error

Spectrum of Sampled Signal

If a signal is under sampled (sampled at a rate below the Nyquist rate), the spectrum consists of overlapping repetitions of the sampled spectrum. Because of the overlapping tails a single repetition of the spectrum no longer has the complete information about the unsampled signal, and it is no longer possible to recover it from the sampled signal. To recover the original signal at the receiving end the sampled signal is passed through a lowpass filter with a cut off of $f_s/2$, we get a spectrum that is not the sampled signal but is a different version due to:

- Loss of the tail of the sampled signal spectrum beyond $f_s/2$

- This same tail appears inverted, or folded, onto the spectrum at the cut-off frequency.

This tail inversion is known as aliasing, (or spectral folding or foldover distortion).

The aliasing distortion can be eliminated by cutting the tail (i.e. filtering) of the sampled signal beyond $f > f_s/2$ before the signal is sampled. By so doing, the overlap of successive cycles in the sampled signal is avoided. The only error in the recovery of the unsampled signal is that caused by the missing tail above $f_s/2$.

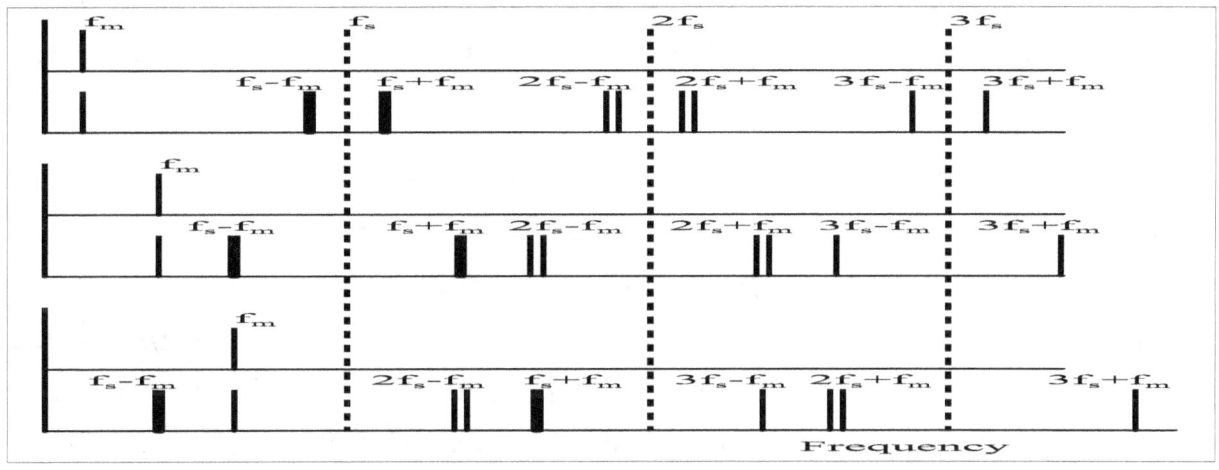

It is simpler to consider aliasing by considering a single frequency component of m(t). We will look at the frequency f_m and it is sampled at a rate f_s. The diagrams show the frequencies which will be present in the sampled signal. There will be frequency components at f_m, $f_s - f_m$, $f_s + f_m$, $2f_s - f_m$, $2f_s + f_m$, $3f_s - f_m$, $3f_s + f_m$, etc. etc.

In the first case f_m is very much less than f_s, so that $f_s - f_m$ is much higher than the cut off of the filter ($f_s/2$).

In the second case f_m is below, but close to $f_s/2$, so that a sharp cut off filter is required to ensure that f_m is passed but $f_s - f_m$ is stopped.

In the third case f_m is higher than $f_s/2$, so that $f_s - f_m$ is less than $f_s/2$. The low pass filter with a cutoff of $f_s/2$ will therefore block f_m (the actual signal frequency) but will pass a signal with frequency $f_s - f_m$.

This is **aliasing:** Strictly speaking, a band limited signal does not exist in reality. It can be shown that if a signal is time limited it cannot be band limited. All physical signals are necessarily time limited because they begin at some finite instant and must terminate at some other finite instant. Hence, all practical signals are theoretically non band limited.

A real signal contains a finite amount of energy; therefore, its frequency spectrum must decay at higher frequencies. Most of the signal energy resides in a finite band, and the spectrum at higher frequencies contributes little. The error introduced by cutting off the tail beyond a certain frequency B can be made negligible by making B sufficiently large.

Thus, for all practical purposes a signal can be considered to be essentially band limited at some value B, the choice of which depends upon the accuracy desired. A practical example of this is a speech signal. Theoretically, a speech signal, being a finite time signal, has an infinite bandwidth. But frequency components beyond 3400 Hz contribute a small fraction of the total energy. When speech signals are transmitted by PCM they are first passed through a lowpass filter of bandwidth of

3500 Hz. (This filter is called an **anti-aliasing filter**). Higher sampling rates (i.e. 8000 samples/sec) permits recovery of the signal from its samples using relatively simple filters i.e. it allows for guard bands between the repetitions of the spectrum (otherwise recovering signals sampled at the Nyquist rate would require very sharp cut-off (ideal) filters).

In summary, aliasing distortion produces frequency components in the desired frequency band that did not exist in the original waveform. Aliasing problems are not confined to speech digitisation processes. The potential for aliasing is present in any sample data system.

Motion picture taking, for example, is another sampling system that can produce aliasing. A common example occurs when filming a rotating wheel. Often the sampling process (the picture refresh rate) is too slow to keep up with the wheel movements and spurious rotational rates are produced. If the wheel rotates 355^0 between frames, it looks to the eye as if it has moved backwards 5^0.

- Voice---is analog in character and moves in the form of waves.
- 3-important wave-characteristics:
    - Amplitude
    - Frequency
    - Phase

Voice Digitization in the Plain Old Telephone Systems (POTS)

- Traditional POTS was analog.
- The current telephone system of the POTS combines both analog and digital transmission technologies.
- Why Voice digitization:
    - Ensures better quality (than analog)

- Provides higher capacity (than analog)
- Deals with longer distance (than analog)

Voice Digitization—How!

- Digitization is just a discrete electrical voltage.
- Electrical pulses can be varied to represent characteristics of an analog voice signal.
- **5-different VD-techniques:**
  - PAM = pulse amplitude modulation
  - PDM = pulse duration modulation
  - PPM = Pulse position modulation
  - PCM = Pulse code modulation
  - ADPCM = Adaptive differential PCM

Analogue-to-digital conversion

an analogue electrical waveform or signal is, by definition, an analogue of an input waveform, i.e. the air pressure variation caused by a speaker. An analogue waveform also has another defining attribute, namely, that it
may take any value within the permitted range. In contrast a digital signal is a series of numerical representations of an input waveform and, importantly, the digital signal can take only a limited number of values, i.e. it has a fixed repertoire – in the case of a binary digital signal this number is two. (It is the use of binary values that enables digital transmission to achieve such good quality performance irrespective of distance, since the signal can be periodically cleaned up and regenerated by comparing the impaired signal with just one threshold, unlike the analogue signal in which impairments accumulate with distance and cannot be corrected.)

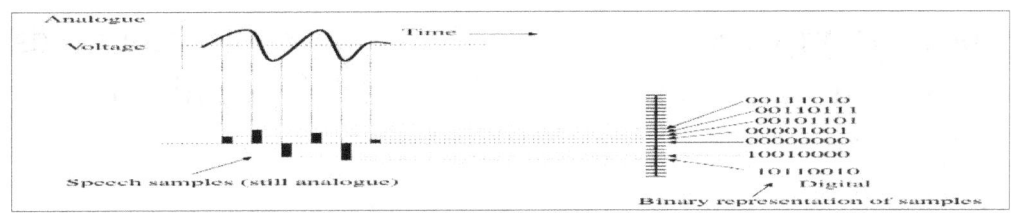

frequency of the waveform, as described earlier for the process of TDM. The waveform is then sampled at or above the Nyquist rate to produce a series of analogue samples or pulses, whose heights represent the waveform at the sampling instants. (This stream of pulses is known as 'pulse amplitude modulation', PAM.) These analogue samples are then converted to digital by the comparison of each sample with a digitalizing ruler. The digital representation is the numerical value of the graduation on the ruler at or just above the PAM pulse height, process generates a binary number for each PAM pulse, the digital output of the D/A converter. Conversion back to analogue from digital operates in reverse to the D/A described earlier. However, in converting each of the binary numbers to the PAM height given by the graduation on the ruler (i.e. decoding) there is an inherent error introduced when the regenerated PAM sample is compared to the sent original PAM sample. This is because of the finite number of graduations on the ruler, the binary number being the next closest reading to the input PAM sample in the A/D process, yet the binary number will generate a PAM sample exactly at the graduation level on the ruler during the D/A process. The amount of difference between the sent and regenerated PAM sample clearly varies between zero and half the distance between ruler graduations. This is known as the quantization error, and it is perceived by a listener as quantization noise.
In practical systems, the quantization noise is kept acceptably low by using a large number of steps, or quantization levels, on the ruler. In addition, the quantization noise is kept proportional to the level of input signal by bunching the levels closer at the lower end of the ruler and wider towards the top – following a logarithmic progression. The A/D process, described earlier, is performed by a digital coder and the D/A

process is provided by a decoder; often the equipment needs to deal with the Go and Return direction, and so a combined digital coder/decoder or 'codec' is used.

The complete A/D-and-D/A process for a telephone network, which is known as 'pulse-code modulation' (PCM),. This figure plots the various

representations of the signal, from analogue waveform, to PAM samples, to digital numbers or PCM 'words' and back. Notice, that once converted to digital (PCM) the signal may pass through any number of digital transmission systems and switching exchanges across the network.

The A/D process is used for a number of applications in everyday life in addition to telephone networks. consider how the trend towards the use of digital in fixed and mobile telecommunications, entertainment, information and computing creates possibilities for convergence and what this means for the next generation of networks.

2.2 Pulse Amplitude Modulation

Pulse Amplitude Modulation

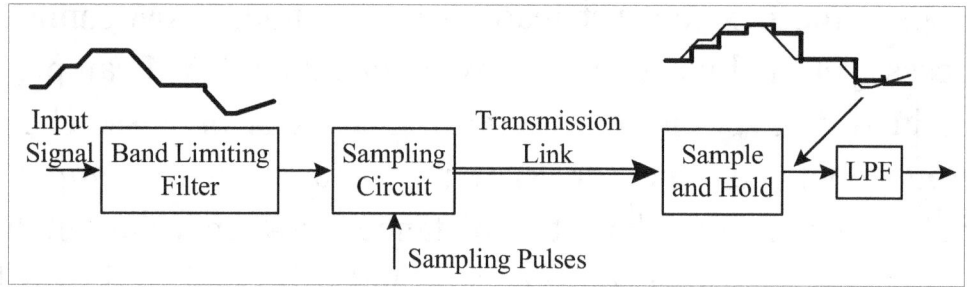

A sampled signal consists of a train of pulses, where each pulse corresponds to the amplitude of the signal at the corresponding sampling time. The signal sent to line is modulated in amplitude and hence the name **Pulse Amplitude Modulation** (PAM).

A complete PAM system must include a band limiting (or **anti-aliasing**) filter before sampling to ensure that no spurious or source-related signals get folded back into the desired signal bandwidth - no aliasing. The input filter may also be designed to cut off very low frequencies to removed 50 Hz hum from power lines.

Several PAM signals can be multiplexed together as long as they are kept distinct and are recoverable at the receiving end. This system is one example of Time Division Multiplex (TDM) transmission (although it has never been widely used fo speech, it has applications in remote monitoring and telemetry).

The sample-and-hold circuit takes in each pulse and holds the amplitude of that pulse until the arrival of the next pulse. It produces a staircase approximation to the sampled wave form. With use of the staircase approximation, the power level of the signal coming out of the reconstructive filter (LPF) is nearly the same as the level of the sampled input signal.

The filters are assumed to have ideal characteristics - not like real filters. Filters with real attenuation slopes at the band edge can be used if the input signal is slightly over sampled. If sampling frequency is greater than twice the bandwidth, the spectral bands are sufficiently separated from each other that filters with gradual roll-off characteristics can be used.

As an example, sampled voice systems typically use band limiting filters with a 3 dB cut-off around 3.4 kHz and a sampling rate of 8 kHz. Thus the sampled signal is sufficiently attenuated at of 4 kHz to adequately reduce the energy level of the foldover spectrum.

- First step in digitizing an analog waveform
- Establishes a set of discrete times at which the input signal waveform is sampled.
- The sampling process is equivalent to amplitude modulation of a constant amplitude pulse train, thus, PAM.

- Nyquist Sampling Rate: The minimum sampling frequency
  - required to extract all information in a continuous, time varying waveform.
  - Nyquist Criterion: $f_s > 2 \times BW$
    f_s: sampling rate, BW: bandwidth of the input signal

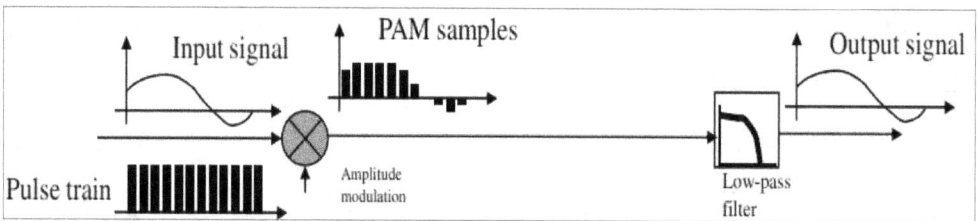

Spectrum of PAM Signal:

The PAM spectrum can be derived by observing that a continuous train of impulses has a frequency spectrum consisting of discrete terms at multiples of the sampling frequency.

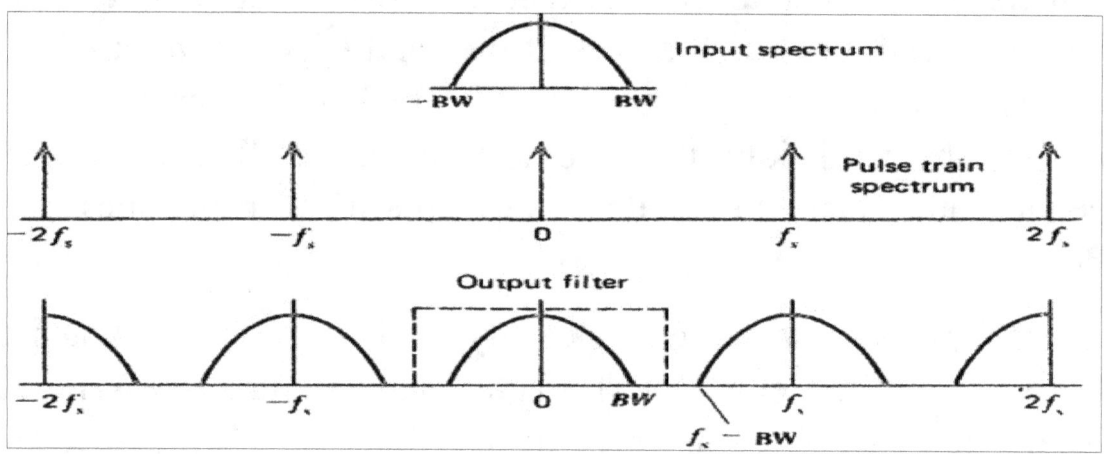

Foldover Distortion:

- If the input is under sampled ($f_s < 2BW$), the original waveform cannot be recovered without distortion
- Another term for this impairment is "aliasing". Remember the old Westerns in which the wheels of stagecoaches appear to move backward.

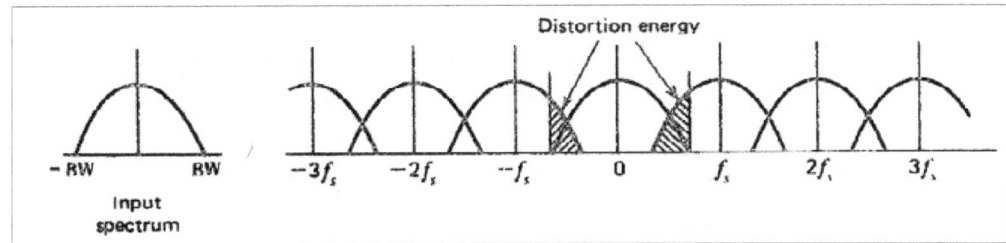

Aliasing:

- The example below demonstrates an aliasing process occurring in speech if a 5.5 kHz signal is sampled at an 8- KHz rate. Notice that the sample values are identical to those obtained from a 2.5 KHz signal.

- Thus, after the sampled signal passes through the 4 KHz output filter, a 2.5 KHz signal arises that did not come from the source.

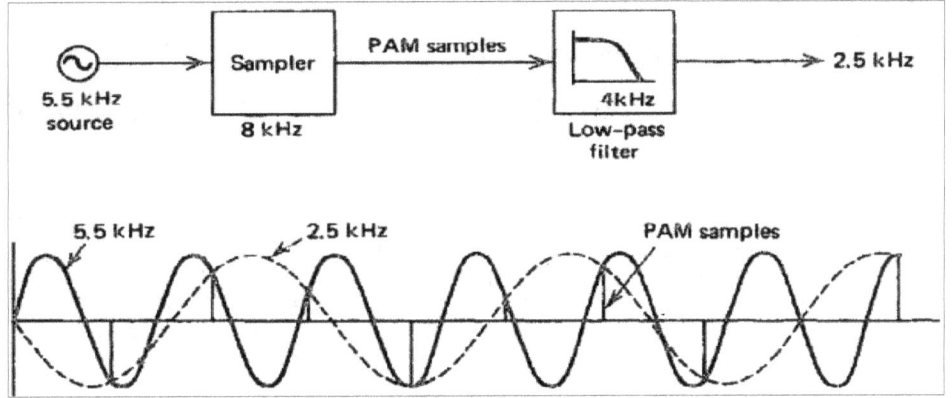

PAM System:

- Complete PAM system includes a band limiting filter before sampling to ensure that no source related signals get folded back into the desired signal bandwidth.

- End-to-End PAM system

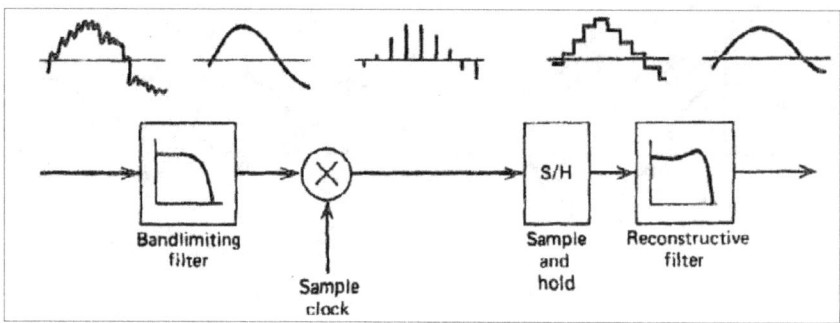

2.3 Pulse Coded Modulation (PCM):

Pulse Code Modulation

Pulse Code Modulation (PCM) is an extension of PAM wherein each analogue sample value is quantized into a discrete value for representation as a digital code word.

Thus, as shown below, a PAM system can be converted into a PCM system by adding a suitable analogue-to-digital (A/D) converter at the source and a digital-to-analogue (D/A) converter at the destination.

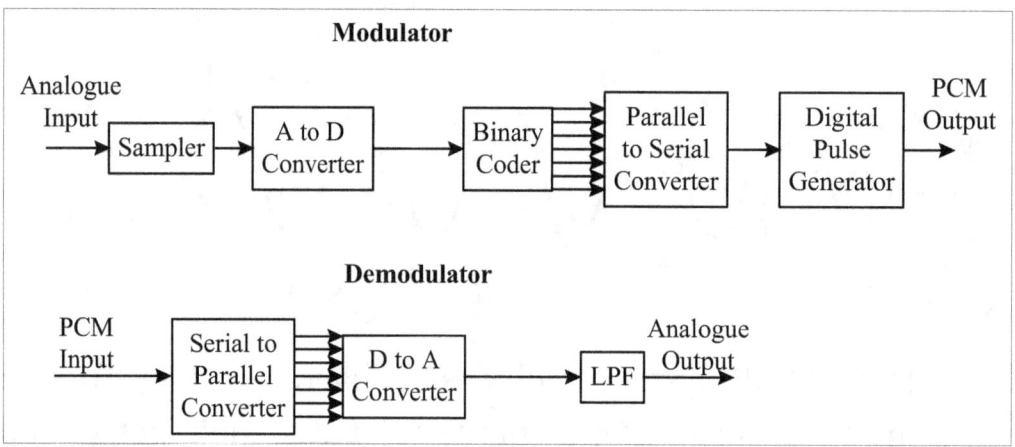

PCM is a true digital process as compared to PAM. In PCM the speech signal is converted from analogue to digital form.

PCM is standardised for telephony by the ITU-T (International Telecommunications Union - Telecoms, a branch of the UN), in a series of recommendations called the G series. For example, the ITU-T recommendations for out-of-band signal rejection in PCM voice coders require that 14 dB of attenuation is provided at 4 kHz. Also, the ITU-T transmission quality specification for telephony terminals require that the frequency response of the handset microphone has a sharp roll-off from 3.4 kHz.

In quantization the levels are assigned a binary codeword. All sample values falling between two quantization levels are considered to be located at the centre of the quantization interval. In this manner the quantization process introduces a certain amount of error or distortion into the signal samples. This error known as quantization noise, is minimised by establishing a large number of small quantization intervals. Of course, as the number of quantization intervals increase, so must the number or bits increase to uniquely identify the quantization intervals. For example, if an analogue voltage level is to be converted to a digital system with 8 discrete levels or quantization steps three bits are required. In the ITU-T version there are 256 quantization steps, 128 positive and 128 negatives, requiring 8 bits. A positive level is represented by having bit 8 (MSB) at 0, and for a negative level the MSB is 1.

Sampling Theorem

The sampling theorem is used to determine the minimum rate at which an analog signal can be sampled without information being lost when the original signal is recovered.

The sampling frequency (f_A) must be more than twice the highest frequency contained in the analog signal (f_S):

$$f_A > 2\, f_S$$

A sampling frequency (f_A) of 8000 Hz has been specified internationally for the frequency band (300 Hz to 3400 Hz) used in telephone systems, i.e. the telephone signal is sampled 8000 times per second. The interval between two consecutive samples from the same telephone signal (sampling interval = T_A) is calculated as follows:

$$T_A = \frac{1}{f_A} = \frac{1}{8000Hz} = 125\mu s$$

Next Figure shows how the telephone signal is fed via a low-pass filter to an electronic switch. The low-pass filter limits the frequency band to be transmitted; it suppresses frequencies higher than half the sampling frequency. The electronic switch - driven at the sampling frequency of 8000 Hz - takes samples from the telephone signal once every 125 µs. A pulse amplitude modulated signal is thus obtained at the output of the electronic switch: a PAM signal.

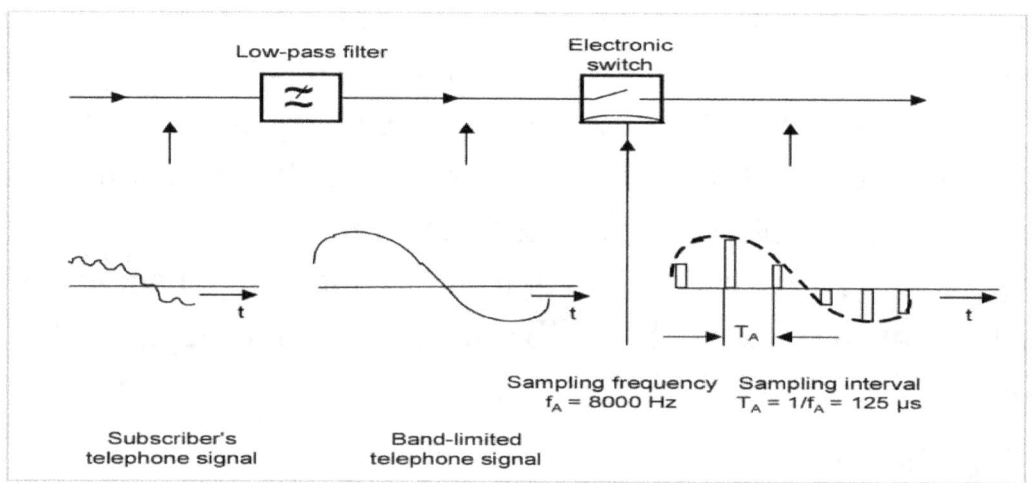

- PCM is an extension of PAM wherein each analog sample is quantized into a discrete value for representation as a digital codeword

- PAM system can be converted to PCM if we add ADC at the source and DAC at the destination.

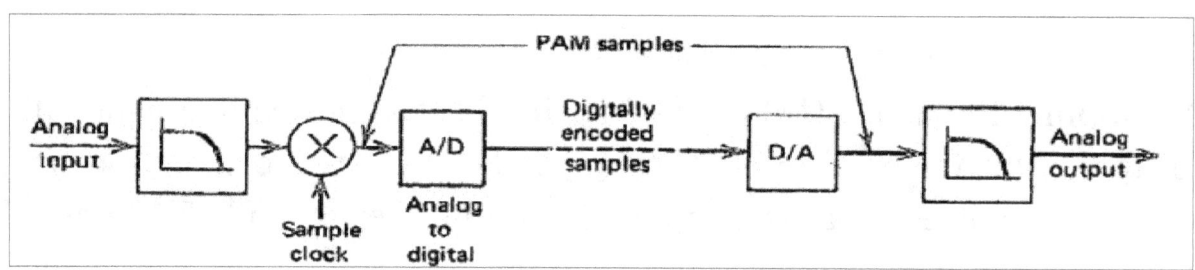

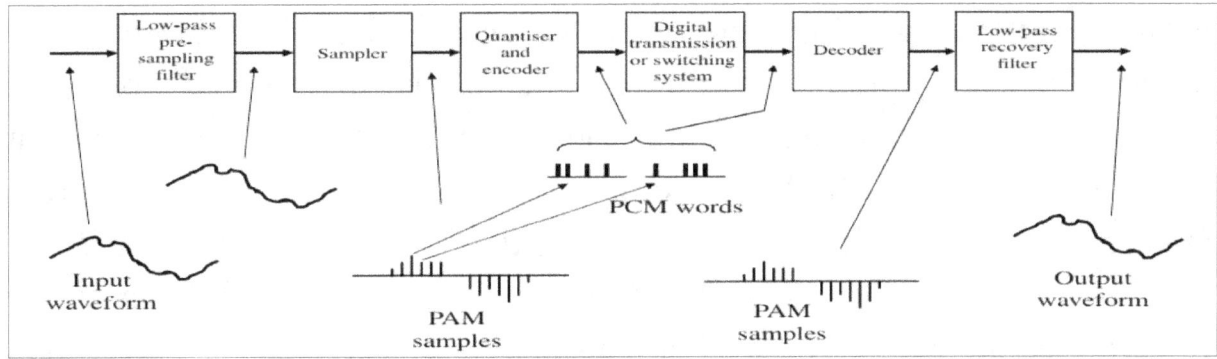

Quantization:

This is the process of setting the sample amplitude, which can be continuously variable to a discrete value. Look at Uniform Quantization first, where the discrete values are evenly spaced.

We assume that the amplitude of the signal m(t) is confined to the range ($-m_p$, $+m_p$). This range ($2m_p$) is divided into L levels, each of step size δ, given by

$$\delta = 2\, m_p / L$$

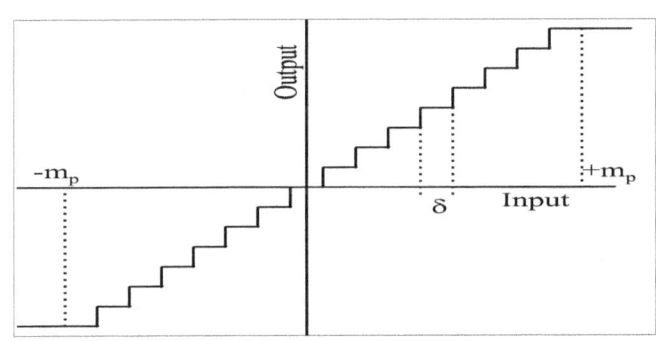

A sample amplitude value is approximated by the midpoint of the interval in which it lies. The input/output characteristics of a uniform quantizer is shown.

The pulse amplitude modulated signal (PAM signal) still represents the telephone signal in analog from. The samples can, however, be transmitted and further processed much more easily in digital form. The first stage in the conversion to a digital signal - in this case a pulse code modulated signal (PCM signal) – is quantizing. The whole range of possible amplitude values is divided into quantizing intervals.

The quantizing principle is shown in next figure. In order to simplify the explanation only 16 equal quantizing intervals are numbered + 1 to + 8 in the positive range of the telephone signal and - 1 to - 8 in the negative range. The appropriate quantizing interval is determined for each sample. Decision values form the boundaries between adjacent quantizing intervals. On the transmit side, therefore, s e v e r a l different analog values fall within the same quantizing interval. On the receive side o n e signal value, corresponding to the midpoint of the quantizing interval, is recovered for each quantizing interval. This causes small discrepancies to occur between the original telephone signal samples on the transmit side and the recovered values. The discrepancy for each sample can be up to half a quantizing interval. The quantizing distortion which may arise on the receive side as a result of this manifests itself as noise superimposed on the useful signal. Quantizing distortion decreases as the number of quantizing intervals are increased. If the quantizing intervals are made sufficiently small the distortion will be minimal and the noise imperceptible.

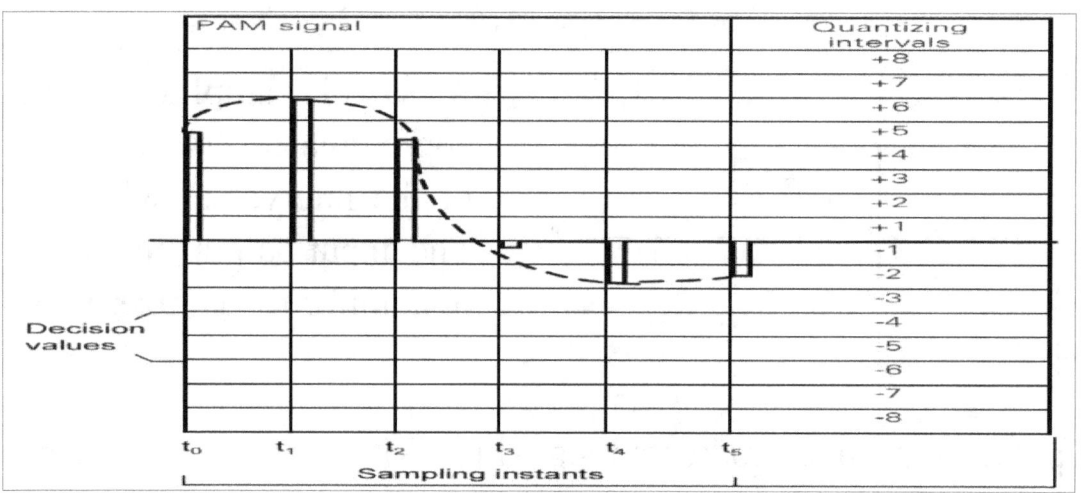

If equally large quantizing intervals are used over the whole amplitude range, relatively large discrepancies will occur in the case of small signal amplitudes (uniform quantizing,). These discrepancies might be of the same order of magnitude as the input signals themselves and the signal-to-quantizing noise ratio would not be large enough. For this reason 256 unequal quantizing intervals are therefore used in the practice (non-uniform quantizing):

- small quantizing intervals for lower signal values

larger quantizing intervals for higher signal values the ratio of the input signal to the possible discrepancy as a result of quantizing is therefore approximately the same for all input signal values. Non-uniform quantizing is specified with the aid of characteristics. The CCITT recommends two such characteristics in G.711:

a) the "13 segment characteristic" (A-law, e.g. for the PCM30 transmission system in Europe)

b) the "15 segment characteristic" (µ-law, e.g. for the PCM24 transmission system in the USA)

- Quantization process has a set of quantization intervals associated in a one-to-one fashion with a binary codeword.

- Binary codeword corresponds to a discrete amplitude

- Quantization process introduces a certain amount of error or distortion into the signal samples.

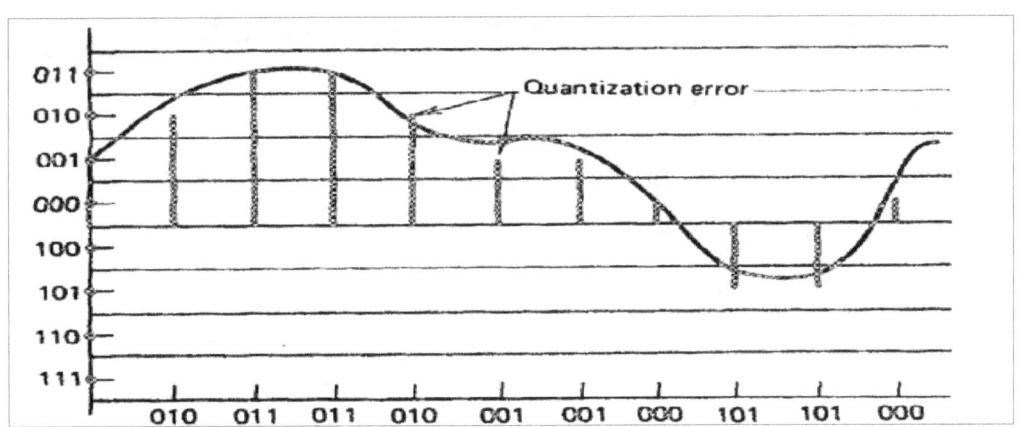

Quantization Noise

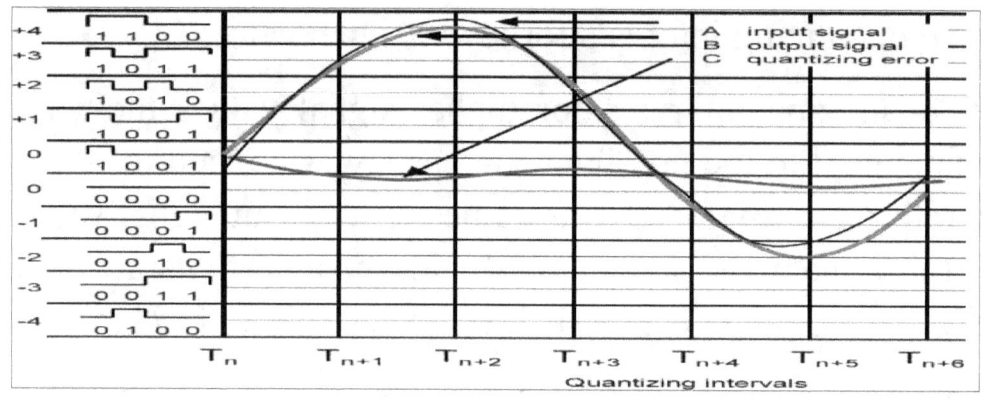

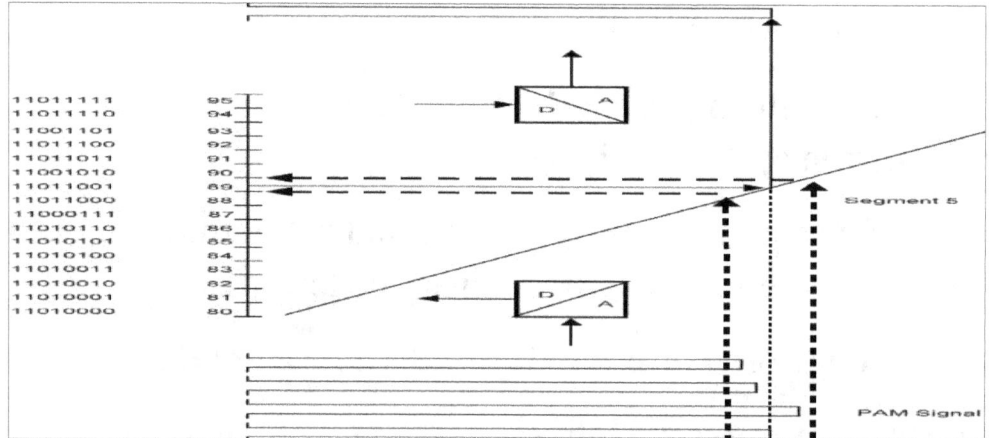

- Quantization errors of a PCM encoder are generally assumed to be distributed randomly and uncorrelated to each other
- Treated as an additive noise
- Signal-to-quantizing-noise ratio (SQR) to define amount of quantization distortion.

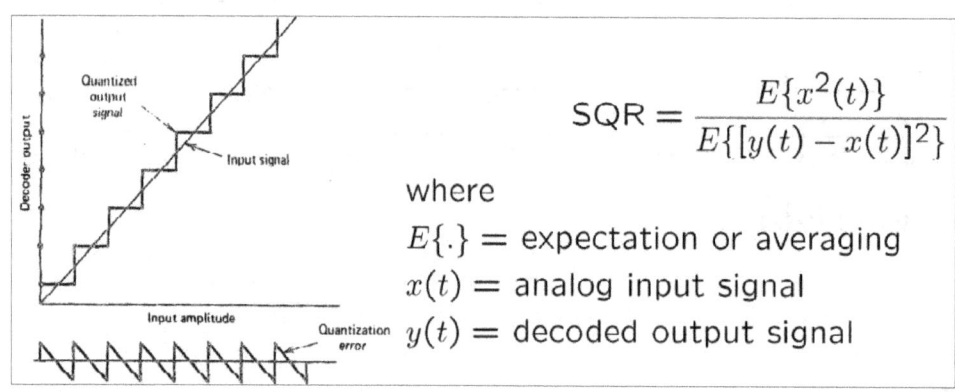

$$SQR = \frac{E\{x^2(t)\}}{E\{[y(t) - x(t)]^2\}}$$

where

$E\{.\}$ = expectation or averaging
$x(t)$ = analog input signal
$y(t)$ = decoded output signal

Quantization Noise

- The error e(t)=y(t)-x(t) is limited in amplitude to q/2, where q is the height of the quantization interval.

- A sample value is equally likely to fall anywhere within a quantization interval implying a uniform probability density of amplitude 1/q.

$$p(\eta) = \begin{cases} \frac{1}{q} & -\frac{q}{2} \leq \eta \leq \frac{q}{2} \\ 0 & \text{otherwise} \end{cases}$$

The average or expected value of noise power is determined as

Quantization noise power $= \int_{-q/2}^{q/2} \left(\frac{1}{q}\right) \eta^2 d\eta = \left(\frac{1}{12}\right) q^2$

$\text{SQR(dB)} = 10 \log_{10} \left(\frac{v^2}{q^2/12}\right) = 10.8 + 20 \log_{10} \left(\frac{v}{q}\right)$ $v \to$ rms amplitude

$\text{SQR(dB)} = 10 \log_{10} \left(\frac{A^2/2}{q^2/12}\right) = 7.78 + 20 \log_{10} \left(\frac{A}{q}\right)$ $A \to$ sinewave amplitude

Idle Channel Noise

- The noise may actually be greater than the signal when sample values are in the first quantization interval.

- Bothersome during speech pauses

- Midriser Quantizer: It cannot generate constant zero output level

- Midtread Quantizer: It straddles the origin to generate zero Output

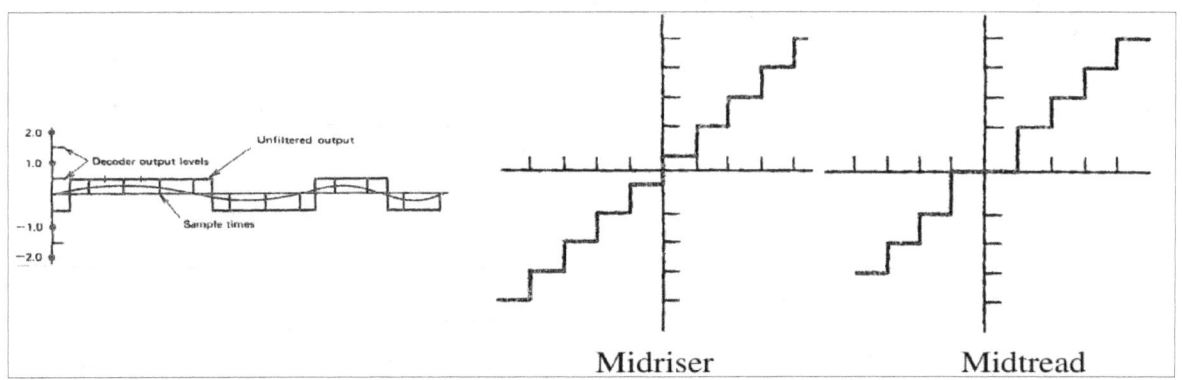

Midriser   Midtread

- Numerical value of each codeword is proportional to th quantized amplitude that it represents
- Minimum digitized voice quality requires 26 dB SQR to provide adequate quality for small signals
    - $q_{max}=0.123A$, and encoding from $-A$ to A, we need 4 bits
- Dynamic Range (DR): Capability of transmitting a large range of signal amplitudes
    - Expressed in dB as the ratio of the maximum amplitude signal to the minimum amplitude signal

$$DR = 10 \log_{10}\left(\frac{P_{max}}{P_{min}}\right) = 20 \log_{10}\left(\frac{V_{max}}{V_{min}}\right)$$

    - PCM performance equation for uniform coding

$$SQR = 1.76 + 6.02n + 20 \log_{10}\left(\frac{A}{A_{max}}\right)$$

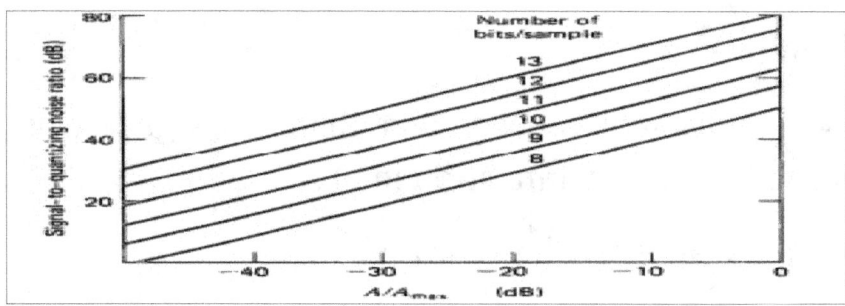

Companding:

In a uniform or linear PCM system the size of every quantization interval is determined by the SQR requirement of the lowest signal to be encoded. This interval is also for the largest signal - which therefore has a much better SQR.

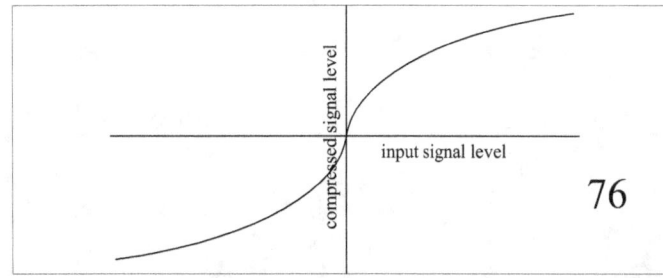

Originally to produce the non linear quantization the baseband signal was passed through a non-linear amplifier with input/output characteristics as shown before the samples were taken. Low level signals were amplified and high-level signals were attenuated. The larger the sample value the more it is **comp**ressed before encoding. The PCM decoder ex**pands** the compressed value using an inverse compression characteristic to recover the original sample value. The two processes are called **companding**.

There are 2 companding schemes to describe the curve above:

1. **μ-Law Companding:** (also called log-PCM) This is used in North America and Japan. It uses a logarithmic compression curve which is ideal in the sense that quantization intervals and hence quantization noise is directly proportional to signal level (and so a constant SQR).

2. **A- Law Companding**: This is the ITU-T standard. It is used in Europe and most of the rest of the world. It is very similar to the μ-Law coding. It is represented by straight line segments to facilitate digital companding.

Originally the nonlinear function was obtained using nonlinear devices such as special diodes. These days in a PCM system the A to D and D to A converters (ADC and DAC) include a companding function.

Example: A 26 dB SQR for small signals and a 30 dB dynamic range produces a 56 dB SQR for the maximum amplitude signal.

In this way a uniform PCM system provides unneeded quality for large signals. In speech the max amplitude signals are the least likely to occur. The code space in a uniform PCM system is very inefficiently utilised.

A more efficient coding is achieved if the quantization intervals increase with the sample value. When the quantization interval is directly proportional to the sample value (assign small quantization intervals to small signals and large intervals to large signals) the SQR is constant for all signal levels. With this technique fewer bits per sample are required to provide a specified SQR for small signals and an adequate dynamic range for large signals (but still with the SQR as for the small signals). The quantization intervals are not constant and there will be a nonlinear relationship between the code words and the values they represent.

- SQR provides the quantization interval for small signals.
- Large signals are also encoded with the same quantization interval
  - 26 dB SQR (small signals) +30 dB (DR)=56 dB SQR for large signals
  - Uniform PCM provides unneeded quality for large signals
  - Large signals least likely to occur
- Companding = Compression + Expansion
  - Compression: More efficient coding procedure is achieved the quantization intervals are NOT uniform but allowed to increase with sample value
  - Relationship becomes nonlinear between codeword and sample value
  - Expansion: Inverse compression is needed at the receiver

Compression

- Various compression-expansion characteristics can be chosen to implement a compandor
- Compression used in North America has the µ-law characteristics

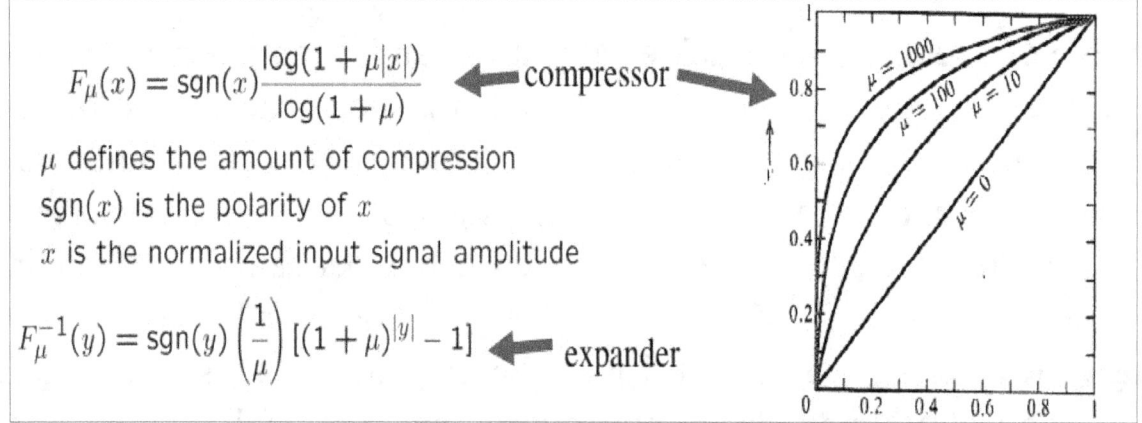

$$F_\mu(x) = \text{sgn}(x)\frac{\log(1+\mu|x|)}{\log(1+\mu)}$$ ← compressor

μ defines the amount of compression
$\text{sgn}(x)$ is the polarity of x
x is the normalized input signal amplitude

$$F_\mu^{-1}(y) = \text{sgn}(y)\left(\frac{1}{\mu}\right)[(1+\mu)^{|y|} - 1]$$ ← expander

Quantizing and Coding for Basic Speech Transmission Systems. The particularities of non-linear quantizing are determined by specific characteristics described in the CCITT-recommendation G.711:

> ➤ The 13-segment characteristic is made up of six linear sections in the positive and negative area. The two segments located at the relative point zero form together a linear segment. Thus, the characteristic comprises a total of 13 segments. In the proximity of point zero there are two Nr. 1 level, a positive and a negative one. The transmission therefore requires in all 2 x 128 = 256 levels. The 13-segment characteristic (also called A-law) is used, for example, for the 30-channel system PCM mainly in Europe.

> ➤ Each quantizing level is allocated an 8-bit code word. The first transmitted bit determines the positive or negative sign of a sample. The following 3 bits ($2^3 = 8$) indicate one of the 7 or 8 segments. The remaining 4 bits ($2^4 = 16$) form the code words for the linear levels within a segment.

> ➤ Systems in accordance with G.711 have a sampling frequency of 8 kHz. Since every 125 µs = 64000 bit/s = 64 kbit/s.

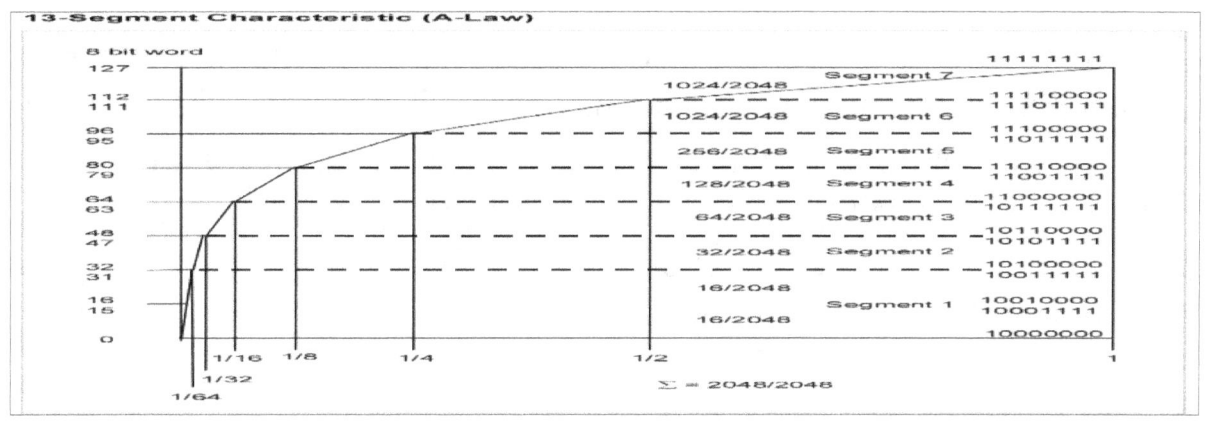

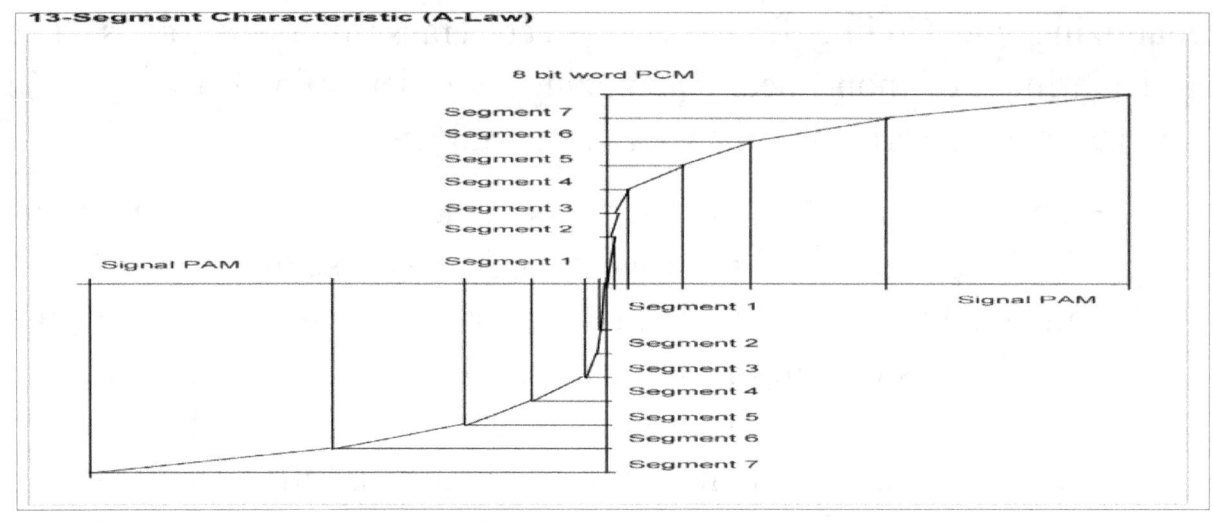

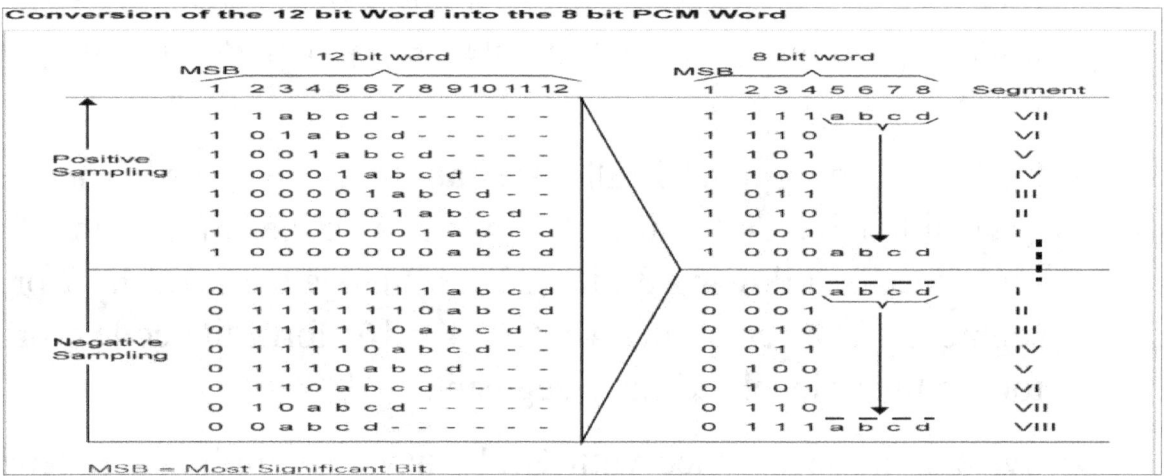

PCM Timing and Synchronisation

The PCM receiver must be able to identify the start and finish of each full sampling sequence and to identify each bit position. The sampling clock needs to be either sent to, or regenerated at, the receiving side to determine when each full sequence of sampling begins and ends. The data clock is also needed to determine exactly when to read each bit of information.

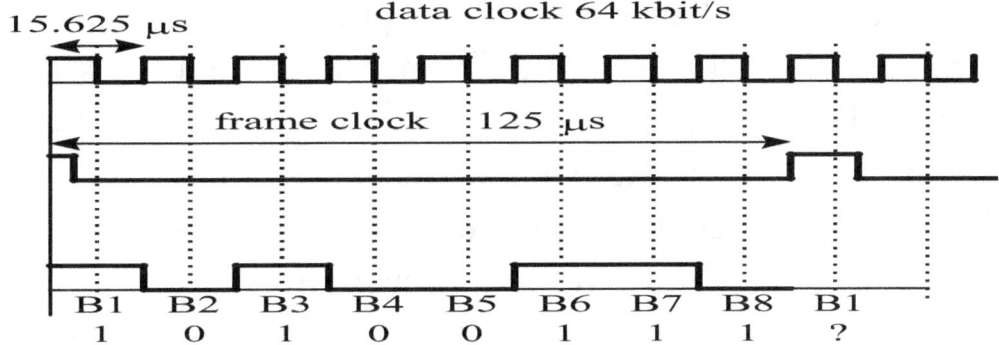

A PCM channel is sampled at 8,000 Hz or once every 125 μs. If there is one channel or 30 TDM channels the sampling period is fixed at 125 μs and this period is known as a frame. Therefore, the frame clock must have a period of 125 μs. The rising edge of the frame clock informs the receiver that the next bit will be Bit 1 of a new sample. The falling edge of the data clock informs the receiver that it must read the data bit.

When the bit stream is transmitted along a line the pulses become distorted and the rise and fall times become significant. Ideally, a 1 will be "high" for 15.625 μs. In practice the pulse may only be above the "high" threshold for a few μs so it is very important that the bit is read within a certain time limit of the clock pulse.

The simplest way to synchronise a PCM sender to a PCM receiver is to send the clock signals on different circuits to the data This would be done in a self-contained system such as private branch exchange (PBX). Telephony is full duplex so that there is a coder and a decoder at each port, but each would use the same clock.

To minimise the number of circuits it is possible to use a line-coding scheme which allows the receiver to extract the clocks from the PCM signal. In this case the receiver will have free running clocks that lock (using a PLL) to the phase and frequency of the transitions in the data stream. The line-coding scheme ensures that there is a transition for every data bit.

2.4 2 Mbit/s Frame and Signaling Pulse Frame

in the direction of transmission the primary multiplexer PCM30 transforms up to 30 signals with different features into 64-kbit/s-digital signals and then combines them by the time division multiplexing procedure to a 2048-kbit/s (2-Mbit/s)-signal, The individual signals can be either LF-speech signals converted by pulse code modulation, or digital signals (e.g. data). In the receive direction a demultiplexer isolates the individual signals out of the 2 Mbit/s signal. The 64-kbit/s-digital signals are then converted again into analog signals. The 2-Mbit/s pulse frame accord. to CCITT-recommendation G.704 consists of 32 time intervals with 8 bits each (octets). In the intervals 1 to 15 and 17 to 31 speech or digital signals are transmitted. Interval 16 contains the channel-associated signaling information (CAS) combined in one multi-frame or, optionally, an additional device specific data channel. In the interval 0 there is an alternate transmission of a frame alignment signal (FAS) or a service word (SVW). In order to isolate the individual signals out of the pulse frame the FAS is searched for in the received 2-Mbit/s-signal. As soon as the bit pattern is recognized, the demultiplexer part of the central multiplexer synchronizes itself to time interval 0. To additionally ensure the synchronization the CRC4-procedure, which will be described in the following, is applied. The service word is used for the transmission of urgent and non-urgent alarms (bit A and bit Sa4), for loop commands (bits Sa6 and Sa7)

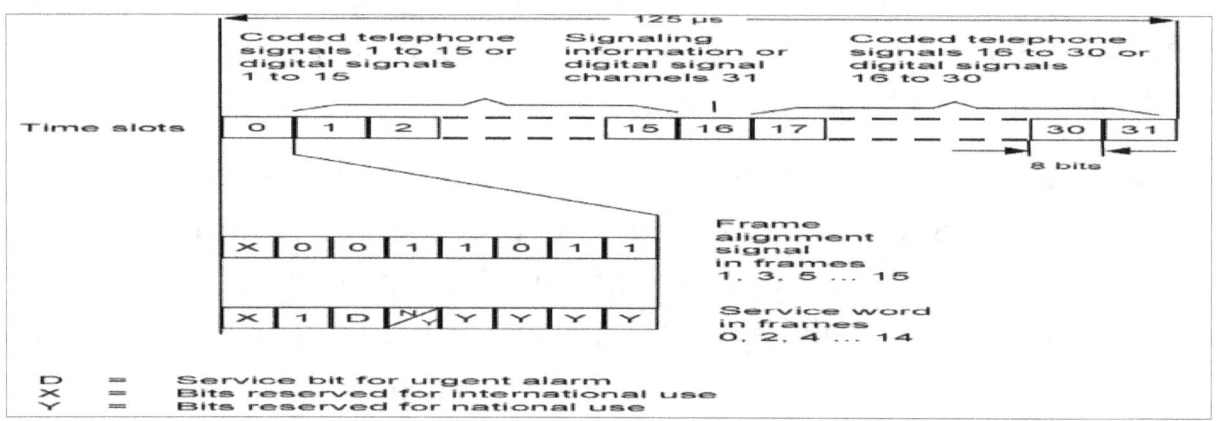

Structure of the Signaling Pulse Frame:

According to CCITT Recommendation G.704 Signaling pulse frame

If analog signal insets are used the PCM30 transmits up to 30 speech signals in the time intervals 1 to 15 and 17 to 31 of the 2-Mbit/s-pulse frame. It has to be ensured that the 64-kbit/s-signals in the time intervals 17 to 31 are counted as channels 16 to 30. The individual channel-associated signaling information is coded with 4 bits (a, b, c, d) separate from the speech signal. The signaling of 30 channels can therefore be combined in 15 octets, which are supplemented by a code and service word of 8 bits, to a multiframe (signaling pulse frame). This multiframe is transmitted in time interval 16 by 16 consecutive 2-Mbit/s-pulse frames (R0 to R15). The code and service word contained in interval R0 is necessary for the multiframe synchronization and for alarm messages.

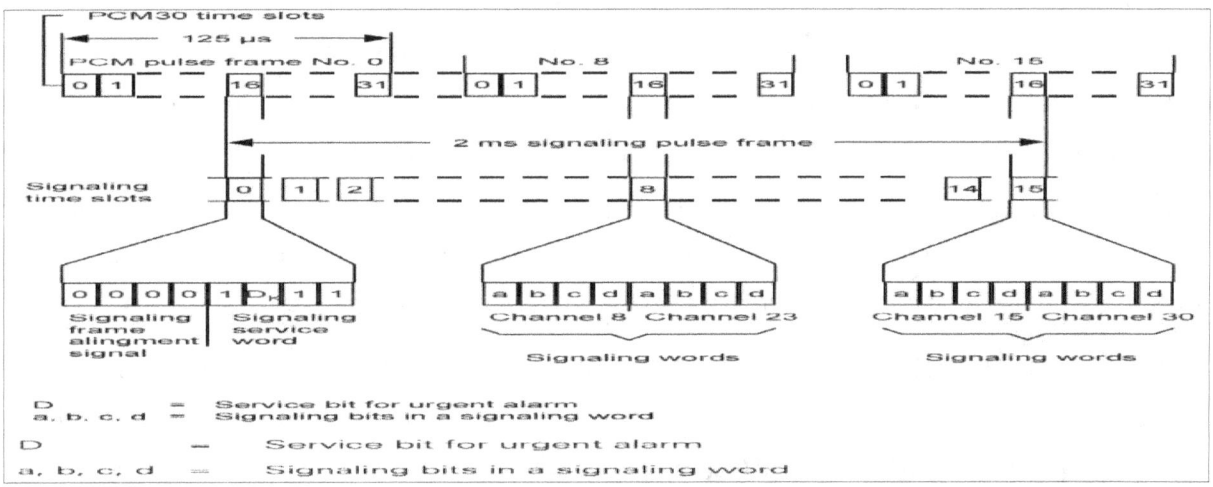

CRC4-Synchronization for Primary Multiplexer:

With the data transmission of synchronous 64 kbit/s digital signals it is possible that the bit patterns of the FAS and the SVW are transmitted (either randomly or on purpose) in the time intervals defined for user signals. If there is a synchronization of the receive side demultiplexer to this bit pattern, an isolation of the individual signals is impossible. Therefore, the CRC4-procedure (Cyclic Redundancy Check by 4 bits) described in CCITT-recommendation G.704 is used in addition, to ensure the synchronization. For this, 16 consecutive 2-Mbit/s frames are combined to a CRC4 multiframe consisting of 2 data blocks and of the multiframe parts I and II. The highest rating bits of the service words in the first twelve 2-Mbit/s frames form the multiframe code word ('001011'). Here, the synchronization is based on two criteria: finding the FAS of the 2-Mbit/s frame and the FAS of a CRC4 multiframe. To continually supervise the synchronization, a data block (e.g. block I) is modified in a data transmitter accord. to a certain algorithm, whereby a rest of 4 bits (the control bits C1 and C4) is left over. These bits are transmitted as highest rating bits in the 2-Mbit frame alignment words of the following data block (block II). The data receiver processes the incoming data block according to the same algorithm as the transmitter. Again, a rest of 4 bits is left over, which are compared individually to the control bits received in the next data block (block II). In case of a correspondence, block I is considered to be error-free. If 915 or more out of 1000 checked blocks were found to be faulty, a new synchronization is started. A CRC4-error is indicated by two E-bits (CCITT-Redbook: Si-bits) at the transmit side; these two E-bits are transmitted as highest rating bits of the service words in the 2-Mbit/s frames 13 and 15 of the CRC4 multiframe. The BER of the 2-Mbit/s signal can be derived from the number of faulty blocks. Thus, for example, a number of 512 or more faulty blocks within a measuring interval of 1 s results in a BER $> 10^{-3}$

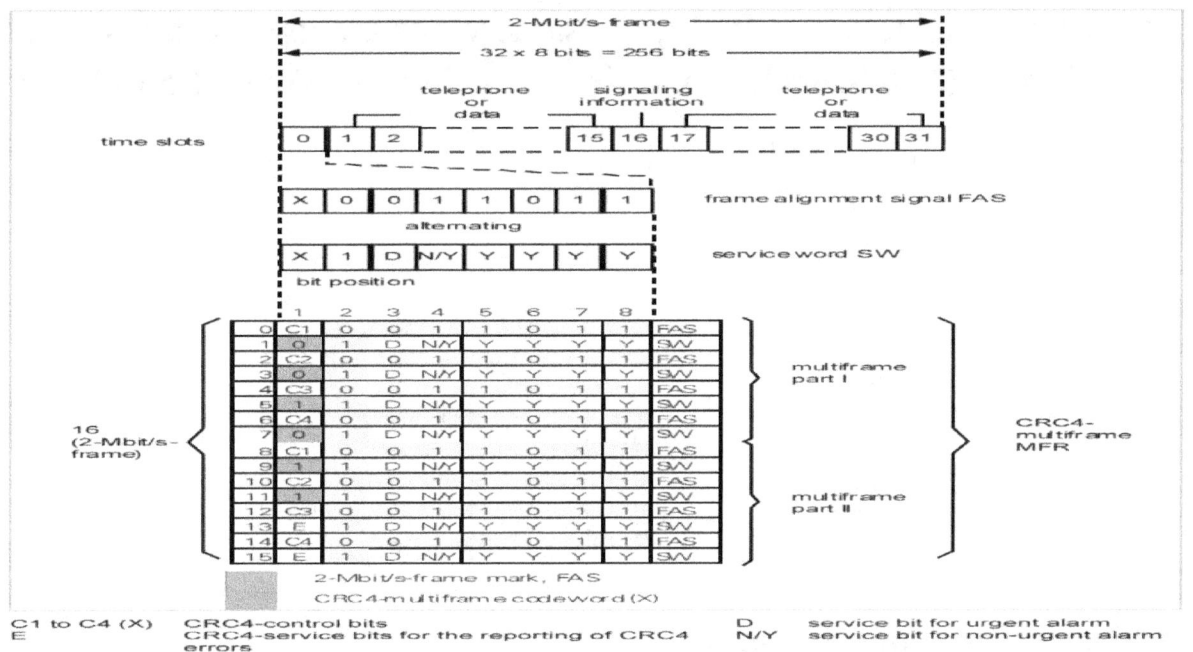

•	Common characteristics	PCM30 and PCM24	
a	Sampling frequency	8 kHz	
b	No. of samples per telephone signal	8.000/s	
c	Pulse frame period	$\frac{1}{b} = \frac{1}{8.000/s} = 125\mu s$	
d	No. of bits in a PCM word	8 bits	
e	Bit rate of a telephone channel	b.d = 8.000/s 8 bits = 64 kbit/s	
•	System-specific characteristics	PCM30	PCM24
f	Encoding/Decoding No. of segments in characteristic	A-law 13	µ-law 15
g	Number of channel time slots per pulse frame	32	24
h	Number of bits per pulse frame (* = additional bit)	d.g = 8 bits. 32 = 256 bits	d.g + 1* = 8 bits. 24 + 1* = 193 bits
i	Period of an 8-bit channel time slot	$\frac{c.d}{h} = \frac{125\mu s.8}{256} =$ aprox. 3.9 µs	$\frac{c.d}{h} = \frac{125\mu s.8}{193} =$ aprox. 5.2 µs
k	Bit rate of time division multiplex signal	b.h = 8.000/s. 256 bits = 2.048 kbit/s	b.h = 8.000/s. 193 bits = 1.544 kbit/s

2.5 T1 Carrier System:

- Telephone wires were used to transmit one audio signal of bandwidth 4 KHz

- After introduction of PCM, the same wires are used to transmit 24 TDM PCM (DS0) telephone signals wit a total BW of 1.544 MHz
- Repeaters every at 6000 feet

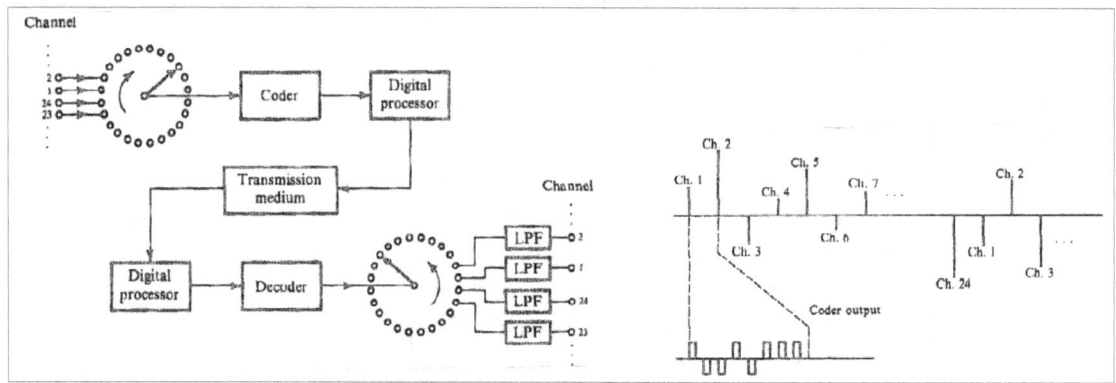

T1 System Signaling Format

- Binary code words corresponding to samples of each of the 24 channels are multiplexed in a sequence
- μ=100, 7 bits are used for data, 1-bit for signaling in D1 signaling
- D2 channel bank needs better voice quality
- 8 bits PCM and μ=255, signaling only repeats every sixth frame

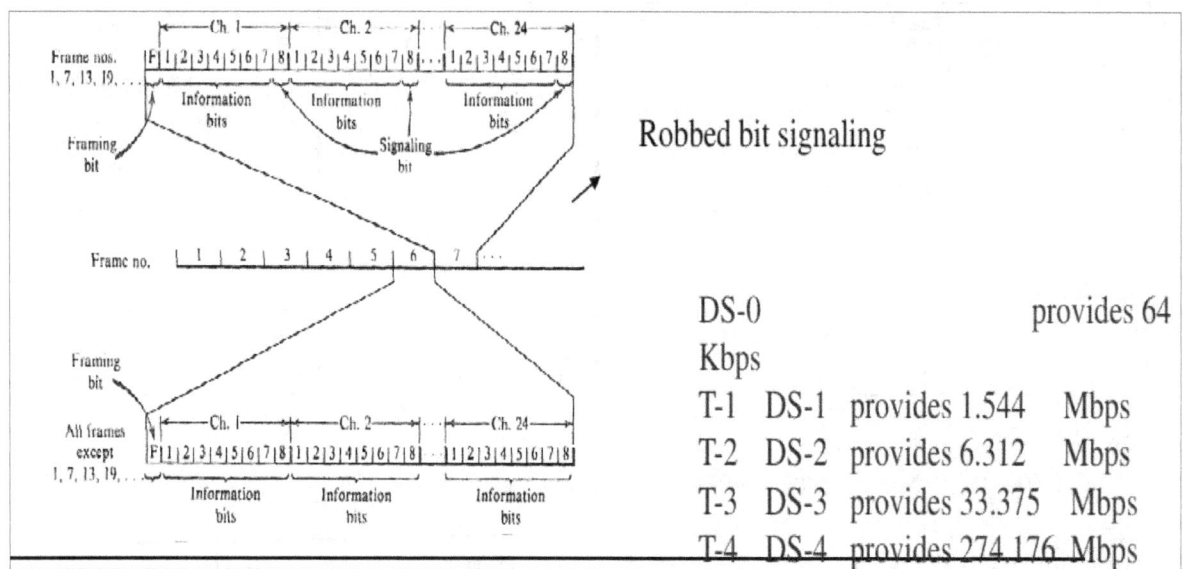

Robbed bit signaling

DS-0 provides 64 Kbps
T-1 DS-1 provides 1.544 Mbps
T-2 DS-2 provides 6.312 Mbps
T-3 DS-3 provides 33.375 Mbps
T-4 DS-4 provides 274.176 Mbps

Voice Compression:

- How does Vonage system work?
- PCM system is inherently capable of encoding an arbitrarily random waveform as long as the maximum-frequency component does not exceed one-half the sampling rate.
- Speech signals have considerable redundancy from one sample to the next.
  - Highly correlated from one sample-to-next, about 85%
  - Ex: Encode only derivative of the signal
- Significant savings in bandwidth are possible through more efficient coding

2.5 Baseband Transmission of Digital Signals and codic

Digital signal devices process the signals as purely binary information, i.e. the signal level does not change between bits with the same logical state. For this reason, these so-called NRZ signals (no return to zero) can only be processed together with the corresponding clock, which enables the identification of individual bit positions. This separate clock is not available for the transmission of data signals and thus it has to be possible to derive (i.e. regenerate) the clock from the data signal on the receiving side. It is obvious that for a NRZ code this is very complicated, if not virtually impossible. A further disadvantage of the NRZ code is that it carries a certain amount of dc-voltage which excludes the signal's galvanic isolation at the interface (transformer etc.). Due to these disadvantages, various interface codes have been developed, all of which comply with the following requirements:

- good clock retrieval features
- no dc-component

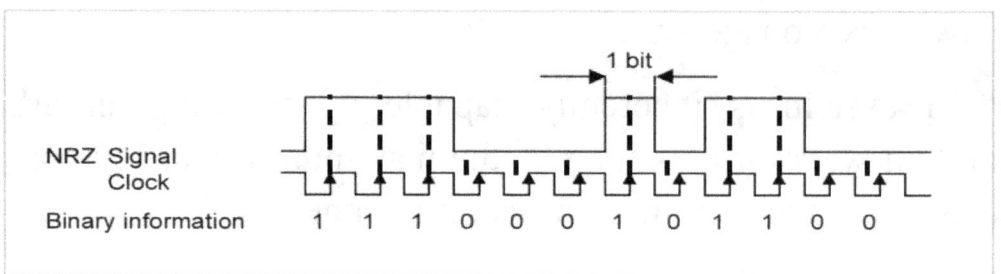

A suitable interface code has a maximum of transitions between the different signal levels, even for the transmission of lengthy sequences of identical logical states; it has no dc-component. The survey shows the development of individual codes.

RZ Code	A log. 1 is represented as half-bit with a change of signals levels from Low → High → Low.	
	Advantage:	clock retrieval possible also for adjacent log. 1 bits.
	Disadvantage:	no clock information for zero sequences, dc-component.
AMI Code	The state log. 1 is represented alternately as positive or negative signal level.	
	Advantage:	clock retrieval possible also for adjacent log. 1 bits, no dc-component.
	Disadvantage:	no clock information for zero sequences
	Is derived from the AMI code. Here, four consequent zero bits are replaced by a 1001 or 0001 combination. This is done in such a way that the signal receiver detects the mutilation of informational contents and cancels it.	
	Advantage:	Maximum clock information, no dc-component.
	Disadvantage:	None
	This code is applied for the device interfaces from 2 Mbit/s up to 34 Mbit/s (baseband transmission). The exact coding rules are enumerated in the following.	
CMI Code	Due to its easy generation with delay lines and simple gate functions the CMI code is suited especially for interfaces with high bitrates. Therefore, this code is standardized for the 140 Mbit/s device interface	

A rather important advantage of the interface code is the possibility it offers to detect transmission errors by supervising the coding rules. With the HDB3 code, for example, the receiving of four zero bits would represent the violation of a coding rule, i.e. at least one-bit error must have been occurred during transmission.

The standardization of interface codes only refers to device interfaces. The codes for conductor-bound transmission paths are manufacturer-dependent and are generally adapted to the requirements of the respective terminating

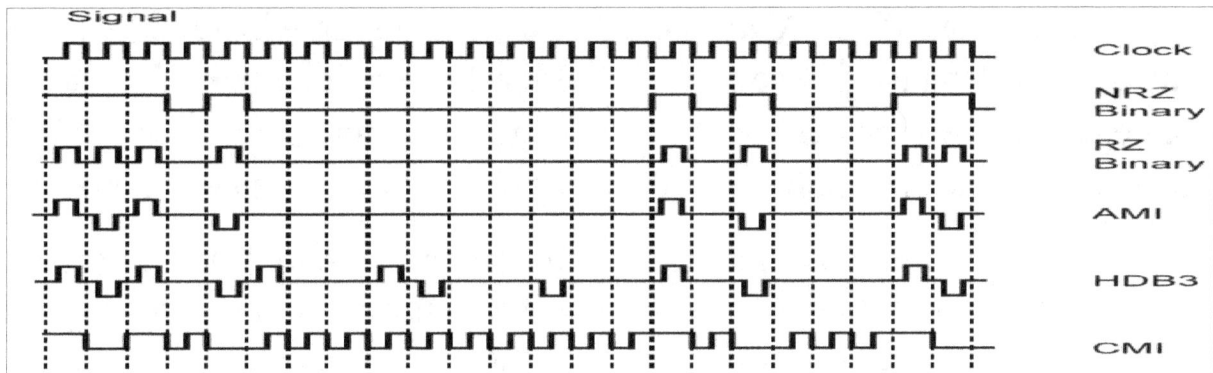

Next Figure shows the amplitude spectrum of various interface codes. For codes without a dc-component the maximum energy is within the range of a frequency which corresponds to half of the bitrate value. This is obvious when comparing the definitions of frequency and bitrate respectively

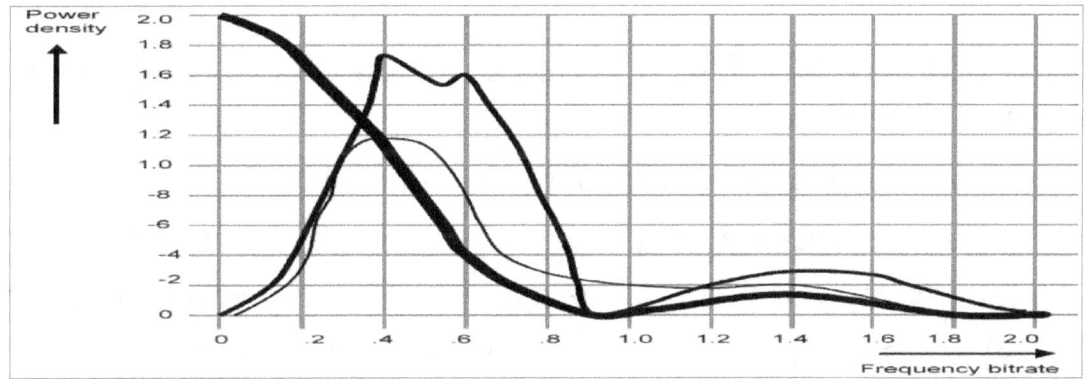

The bit sequence represented in the next figure shall serve as an example. One signal period covers 2 bits and corresponds to the basic wave of the data signal. This wave contains the greatest amount of energy and has a frequency which equals half of the bitrate value. This is also the frequency that is indicated by a frequency counter connected to a source of a digital signal.

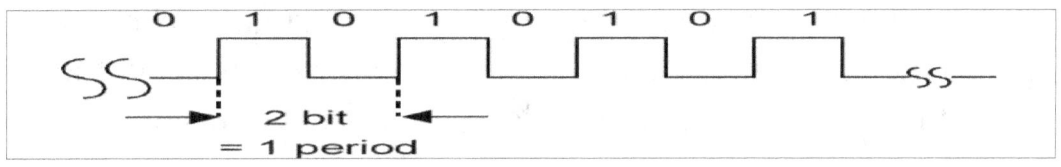

HDB3-Coding rules

(Third-Order-High-Density-Bipolar-Code) The HDB3-code is a modified version of the AMI-code. Binary signals or AMI-code signals may contain lengthy "0" sequences, which hinder the clock retrieval in the regenerative repeaters along digital transmission paths. The HDB3 code enables the elimination of "0" sequences with more than 3 zeros.

The output from the analogue to digital converter (ADC) has n parallel bits. In the case of telephony n = 8. The most significant bit is the signed bit. If the measured sample is positive then the signed bit is 0. If the measured sample is negative then the signed bit is 1. The remaining 7 bits are used to code the sample value. The ITU-T define a look up table which allocates a particular binary code to each quantified A-law value.

The line coding which is used assigns opposite polarities to successive "1" s. This eliminates any DC voltage on the line, and reduces the inter symbol interference if adjacent bits are "1". If there is silence on the PCM channel then the measured samples will be 0 V_{rms} and the output of the DAC will be 1000 0000. A stream of all zeros is not desirable on an active channel because

- all zeros could also be a fault condition and
- it is difficult to recover the clock signal from the incoming signal.

The coding system HDB3 is used and was developed to eliminate all zeros, and to assign opposite polarities to successive "1" s.

This is a bipolar signalling technique (i.e. relies on the transmission of both positive and negative pulses).

In AMI positive and negative pulses (of equal amplitude) are used for alternative symbols 1. No pulse is used for symbol 0. In either case the pulse returns to 0 before the end of the bit interval. This eliminates any DC on the line.

HDB3 encoding rules follow those for AMI, except that a sequence of four consecutive 0's are encoded using a special "violation" bit. The 4th 0 bit is given the same polarity as the last 1-bit which was sent using the AMI encoding rule. This prevents long runs of 0's in the data stream which may otherwise prevent a receiver from tracking the centre of each bit. By introducing violations, extra "edges" are introduced, enabling a Digital PLL to reliably reconstruct the clock signal at the receiver. The HDB3 is transparent to the sequence of bits being transmitted (i.e. whatever data is sent, the Digital PLL can reconstruct the data and extract the bits at the receiver).

To prevent a DC being introduced by excessive runs of zeros any run of more than four zeros encodes as B00V. The value of B is assigned + or - alternately throughout the bit stream.

Example 1 1 1 1 1 1 1 1 = + - + - + - + -

                                                                        B B B B B B B B

            1 0 1 0 1 0 1 0 =   + 0 - 0 + 0 - 0

                                                B 0 B 0 B 0 B 0

            1 0 0 0 0 0 0 1   + 0 0 0 + 0 0 -
                          = B 0 0 0 V 0 0 B

            1 0 0 0 0 1 1 0 = + 0 0 0 + - + 0
                          = B 0 0 0 V B B 0

1. If there are more than 4 consecutive "0"-signal elements, the fourth "0"-signal element shall be replaced by a V-signal element (= "1"-signal element) (000V). Hereby, the V-signal element takes on the same polarity as the "1"-signal element. A V-signal element causes a Violation of the AMI-rule.

2. If between the V-signal element, inserted according to the conditions specified above (rule 1), and the preceding V-signal element there is an even number of "1"-signal elements, then the first of four "0"-signal elements shall be replaced by an A-signal element (= "1"-signal element). The polarity of the A-signal element complies with the AMI rule. The last of four "0"-signal elements becomes again a V-signal

element (A00V). In this case the A- and V-signal elements have the same polarity.

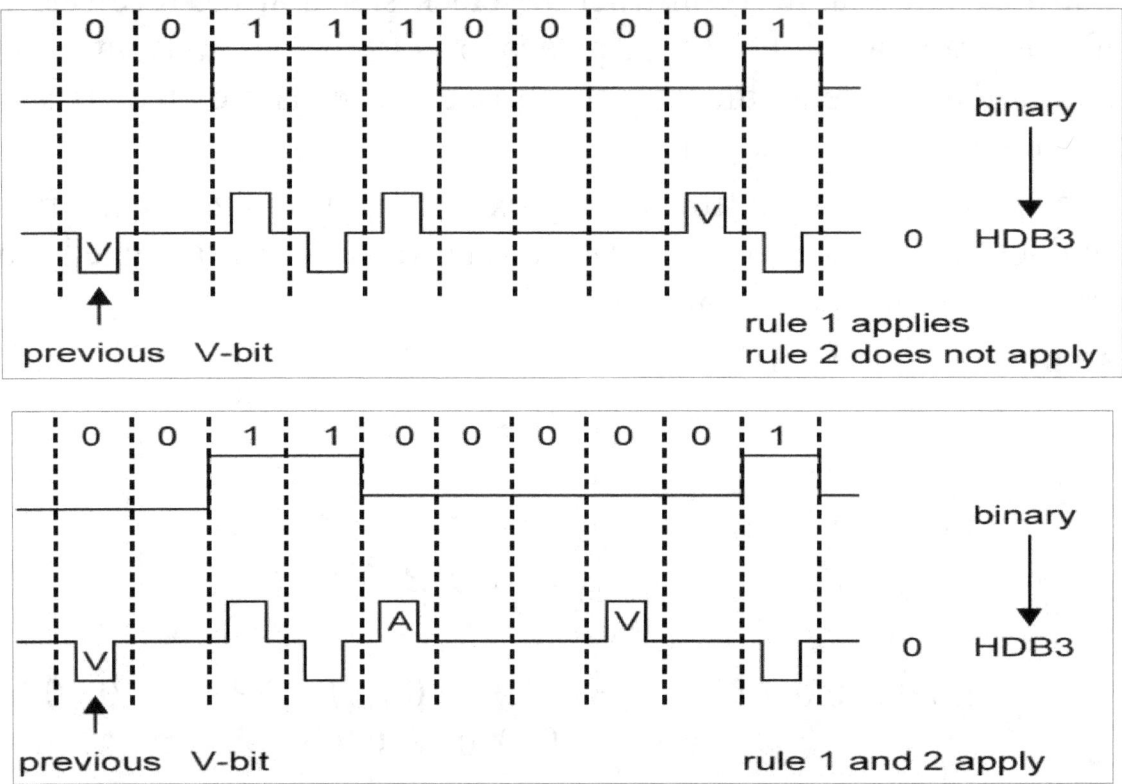

Digital Signal Regeneration:

The digital signal regeneration is one of the advantages of the digital transmission technique. Theoretically, it enables the signals to be transmitted via an unlimited distance without any quality losses. During transmission, a digital signal is attenuated and distorted, which results in a reduction of the signal/noise ratio. The regeneration process has the task of canceling such distortions and regenerate the originally sent signal from the actually received signal. That is why every interface on the receiving side is followed by a regeneration circuit

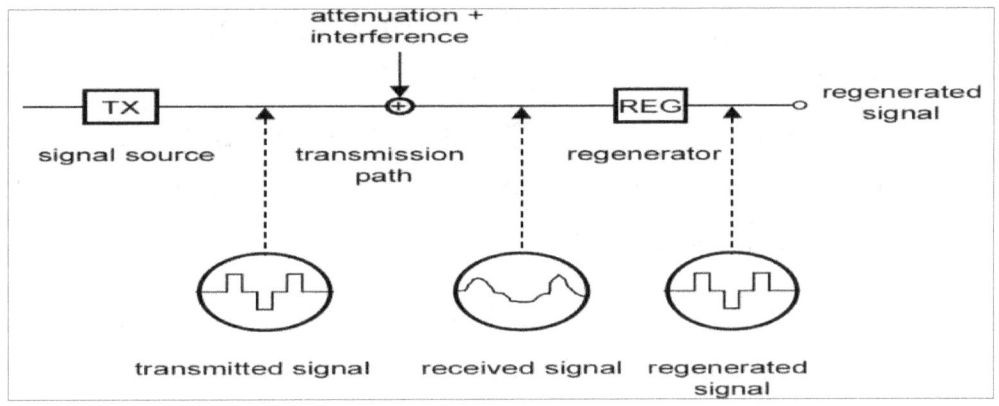

Four basic function blocks are necessary for the digital signal regeneration:

- Amplification block (balancing of attenuation losses)
- Clock retrieval block
- Amplitude decision block
- Time decision block

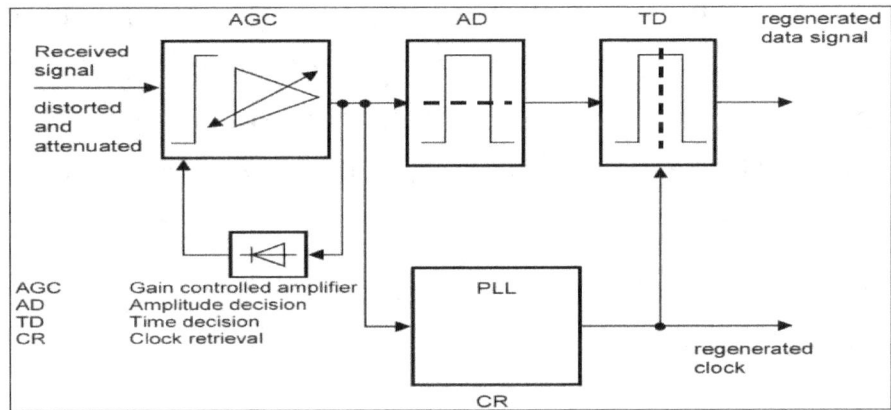

- These four functions are represented in the previous figure: The receiving signal is fed into an automatic gain-controlled amplifier (AGC) which keeps the amplitude of the outgoing signal at a constant value over a wide range of incoming amplitudes. Thus, the attenuation of the transmission path is balanced.

- The constant output level is a precondition for the functioning of the amplitude decision block (AD) which follows. This AD decides on the basis of an internal threshold value whether the level of incoming signal is above or under this threshold value. Accordingly, a signal with the levels Log. 0 or Log. 1 is emitted at the output. The output signal thus consists of pulses, the width of which only depends on the period during which the output signal exceeds the decision threshold.

- The time decision block (TD) has the task of generating signal pulses with constant width. For this, it requires the regenerated receive signal clock which samples the output signal of the amplitude decision block. If, at the time of sampling the signal has a level of Log. 1, the time decision block emits a pulse with constant width. Thus, incoming pulses of any width are turned into pulses corresponding exactly to the bit width of the transmitted signal. The time decision process is the final stage of regeneration.

- The clock retrieval CR block is in charge of regenerating the transmitted signal clock from the receive signal clock. In order to affect this function, a phase locked loop (PLL) is employed, basically consisting of a voltage-controlled oscillator whose frequency can be changed by a control-voltage. By adequate evaluation of the receiving signal it is now possible to reach a control voltage which can set the oscillator to the exact clock frequency value of the transmitting signal.

Reasons for Bit Errors:

The decisive quality criteria for the transmission of digital signals is the bit error rate (BER). This BER represents the proportion of bits which have been mutilated (i.e. incorrectly recorded) during transmission, to the total amount of bits transmitted within a certain interval. The BER directly influences the quality of the transmitted services (e.g. voice channels, data channels, video signals). Two significant BER are explained exemplary in the following:

- **BER = 10-6:** This BER virtually cannot be perceived in a voice channel. For the transmission of data channels, however, this value represents the maximum acceptable limit. The transmission system is in a state of "degraded quality", which is indicated by a degradation alarm (low priority) on the devices involved. The transmission path remains, nevertheless, in operation.

- **BER = 10-3:** This BER causes a strong interference noise in a voice channel. The operating state is judged to be of "unacceptable quality", which is signaled by the devices involved by the emission of a failure alarm (high priority). The transmission path goes out of operation.

How do bit errors arise?

In the previous section it was mentioned that digital signals can be regenerated as requested, i.e. a transmission without quality reduction is possible. This statement is, however, only partially true, i.e. whenever the impairment of the transmission signals is within limits which still permit the regeneration at the receiving side. The reasons for the formation of bit errors are:

- low signal/noise ratio
- jitter
- intersymbol interference
- **Low signal/noise proportion:** Noise amplitudes which influence the amplitude decision process are superimposed to the originally sent signal. The superimposed interference peaks lead to an incorrect signal interpretation at the receiving end. Reasons for a low S/N-ratio are:

  ➢ Too strong signal attenuation during transmission

  ➢ External interference during transmission.

For transmission in cable sections (especially optical fiber) both reasons can be largely eliminated by careful planning

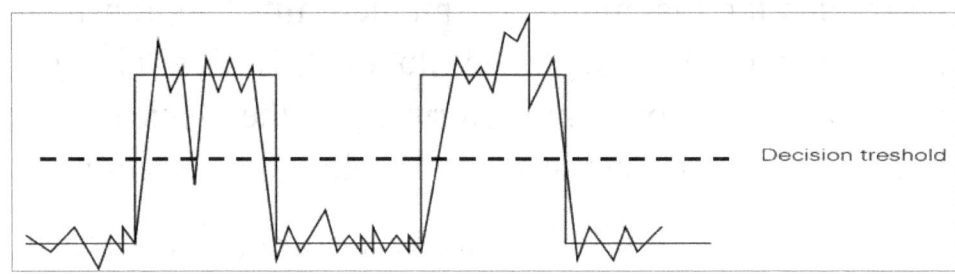

- Jitter: Due to jitter, the transitions between signal levels log. 0 and log. 1 do not take place at periodically recurring points in time (characteristically moments) as for undisturbed signals, which means that the transitions oscillate around the characteristically moments Jitter is characterized by jitter amplitude (unit intervals UI) and jitter frequency. One UI means that, because of deviation from the characteristically moments, the signal edges are within a range equal to the width of 1 bit. The jitter frequency is the number of oscillations around the characteristically moment per one second. Jitter influences the time decision process in the regenerator and causes bit errors for high jitter amplitudes and frequency. Jitter arises in the devices used for signal transmission (i.e. in regenerators and demultiplexers = systematically jitter), or on the transmission path due to external influences (non-systematic jitter).

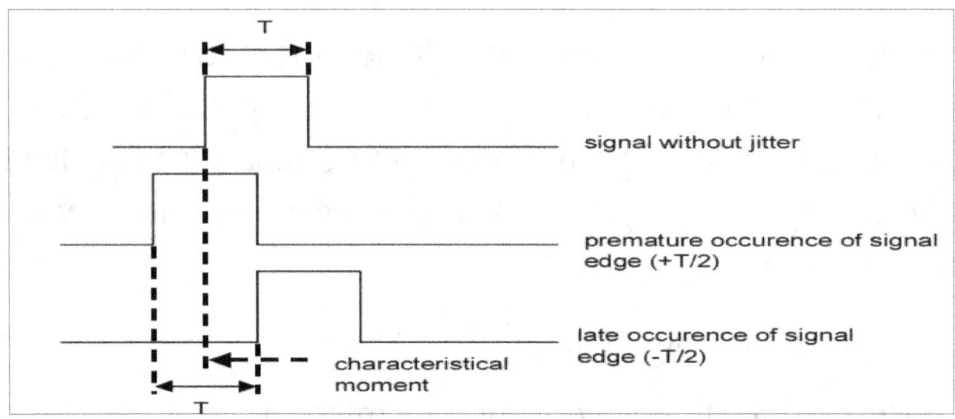

- Inter-symbol interference:

Is caused by a discrepancy between the band width of the transmission path and the bandwidth required for the digital signal. This leads to a bit extension, so that there is an overlap of bits which follow each other. Thus, bit errors occur, the reasons of which can be traced back to the impairment of amplitude decision process. For conductor-bound transmission of digital signals this effect can be excluded by adequate planning. For transmission on radio paths this effect is of fundamental importance as the frequency response of the transmission path can change due to atmospherically influence

Codec :

Codec = Coder + Decoder

- Modem = Modulator + Demodulator
- Codec is a voice-connector device that samples the analog POTS transmission into a stream of binary digits using PCM.
- For quality digital transmission over a Codec:
  - 8000 samples/sec
  - Each sample = 1 byte = 8 bits
  - 8000 X 8 = 64,000 bits/sec----(DS-0 circuit)
  - Uses T-1 circuit—1.544 Mbps---(equivalent to 24 DS-0)

Codec and Modem!

- Codec is a voice-connector device
- Modem is a data-connector device.
- All high-speed Codecs operate at multiples of DS-0 speed:
  - DS-0 provides 64 Kbps
  - T-1 DS-1 provides 1.544 Mbps
  - T-2 DS-2 provides 6.312 Mbps

> T-3 DS-3 provides 33.375 Mbps

> T-4 DS-4 provides 274.176 Mbps

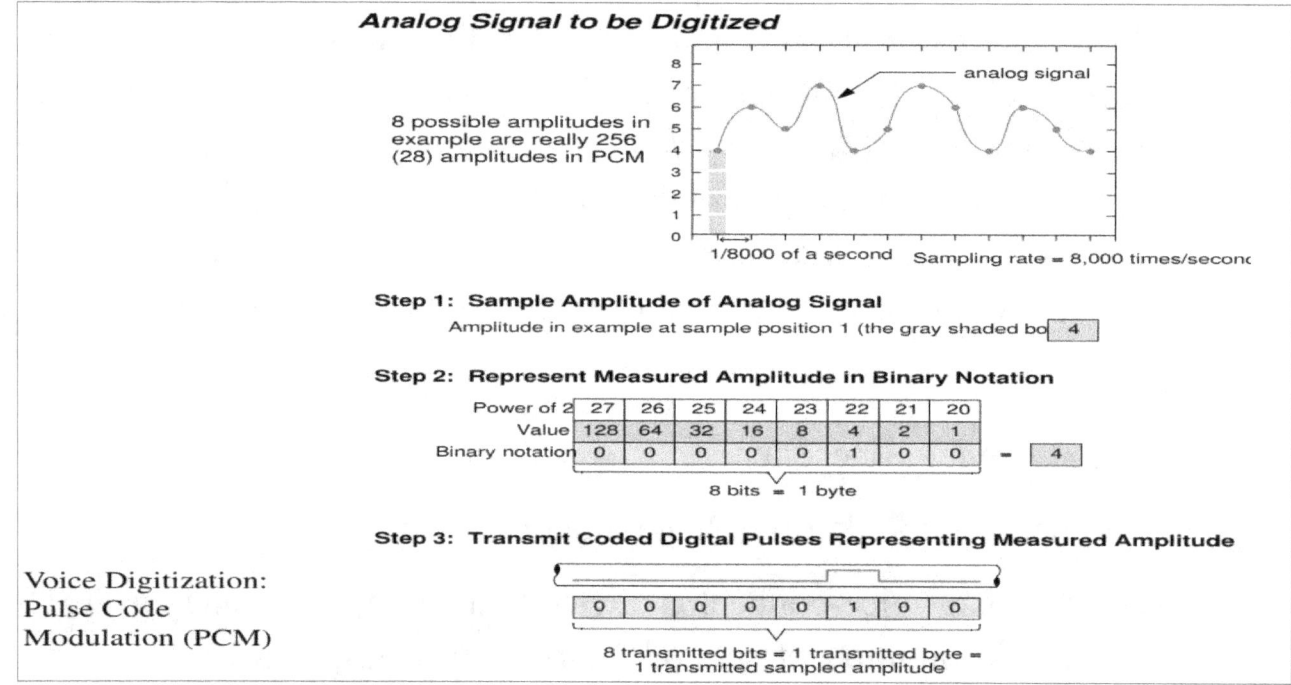

Voice Digitization: Pulse Code Modulation (PCM)

Voice Compression!

- ADPCM--adaptive differential PCM-is also known as a voice compression technique, because of its ability to transmit 24 digitized voice conversations in half the bandwidth required by the PCM.

- More advanced VCs use—digital signal processors that take the digitized PCM and further manipulate and compress it.

- Each voice compression technique seeks to reduce the amount of transmitted voice-info in one way or another.

Differential PCM

- Differential PCM is designed specifically to exploit the sample-to-sample redundancies in a typical speech waveform.

- The range of sample differences is less than the range of individual samples, fewer bits are needed.
- DPCM is a first order prediction How do you predict the future sample?

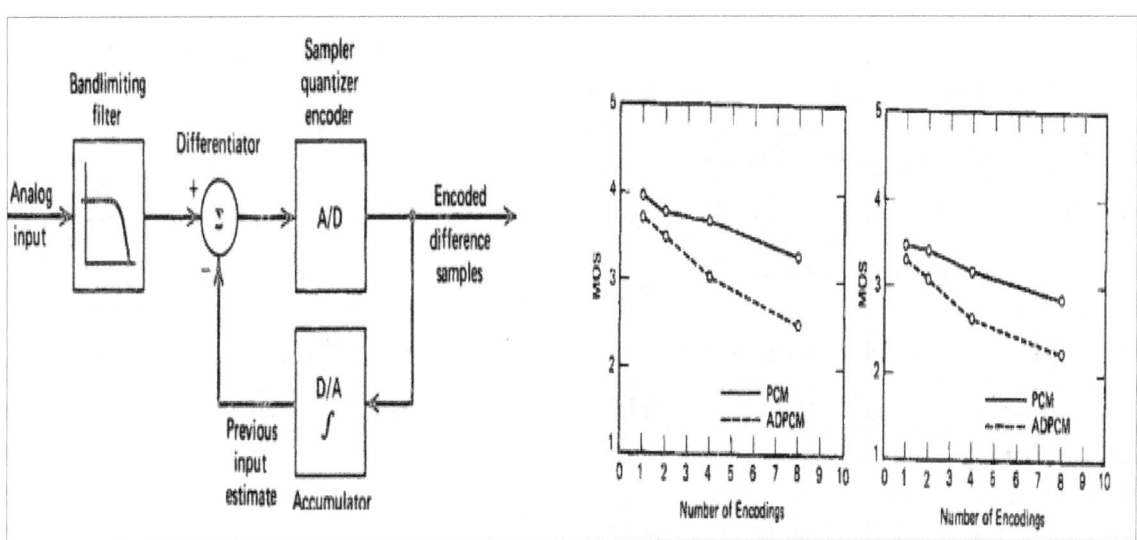

Example:

Speech digitization techniques are sometimes measured for quality by use of an 800-Hz sine wave as a representative test signal. Assuming a uniform PCM system is available to encode the sine wave across a given dynamic range, determine how many bits sample can be saved by using a uniform DPCM system.

- A basic solution can be obtained by determining how much smaller the dynamic range of the difference signal is in comparison to the dynamic range of the signal amplitude. Assume the maximum amplitude of the sine wave is A, so that

$$x(t) = A\sin(2\pi 800 t)$$

- The maximum amplitude of the difference signal can be obtained by differentiating and multiplying by the time interval between samples:

$$\frac{dx}{dt} = A(2\pi)(800)\cos(2\pi 800 t) \quad |\Delta x(t)|_{max} = A(2\pi)(800)\left(\frac{1}{8000}\right) = 0.628A$$

- The savings in bits per sample can be determined as:

$$\log_2\left(\frac{1}{0.628}\right) = 0.67 \text{bits} \longrightarrow \frac{2}{3} \text{ bits less than a regular PCM}$$

Adaptive Differential PCM

- Adaptation logic can be added to the basic DPCM algorithm to create what is referred to as adaptive differential PCM (ADPCM).

- ITU has established a 32-kbps ADPCM standard (G.721) for voice messaging and for increasing the number of voice channels on a T1 line.

- It uses an 8th order predictor, adaptive quantization, and adaptive prediction.

Delta Modulation

- Special case of DPCM with 1-bit per sample difference signal

- Requires high-sampling rate than DPCM

- Problem: Slope overload with sampling rate is low

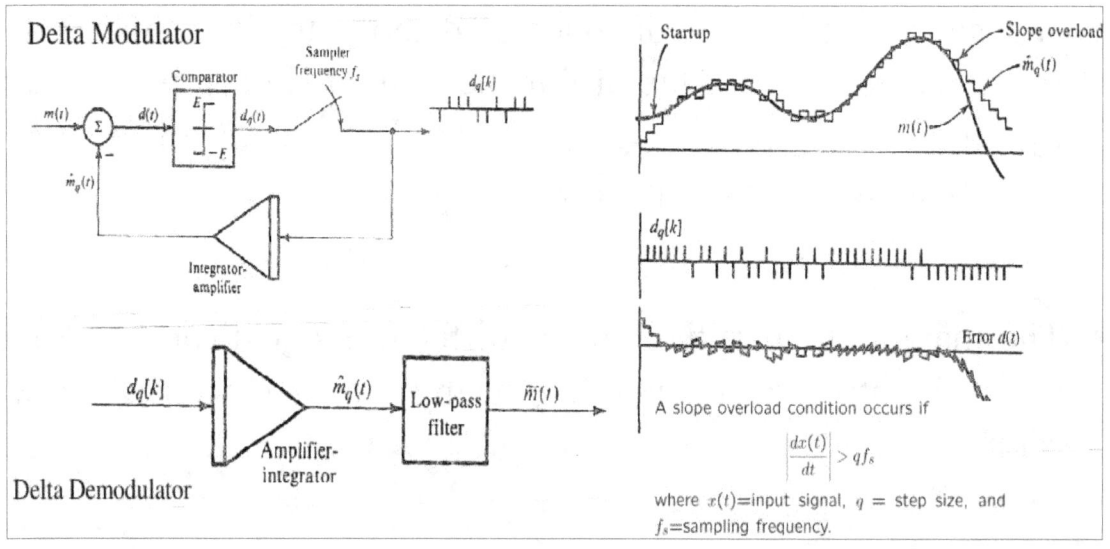

Subband Coding

- Human ear perceives speech by measuring the short-term energy level of individual frequency bands

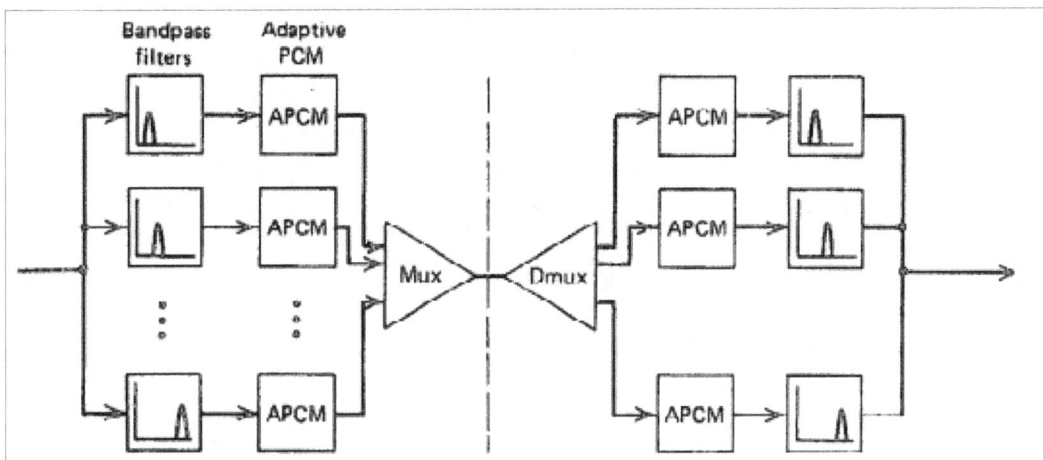

2.6 Digital Transmission (Line Coding):

Pulse Transmission

- Source Multiplexer Line Coder
- Line Coding: Output of the multiplexer (TDM) is coded into electrical pulses or waveforms for the purpose of transmission over the channel (baseband transmission)
- Many possible ways, the simplest line code on-off
- All digital transmission systems are design around some particular form of pulse response.

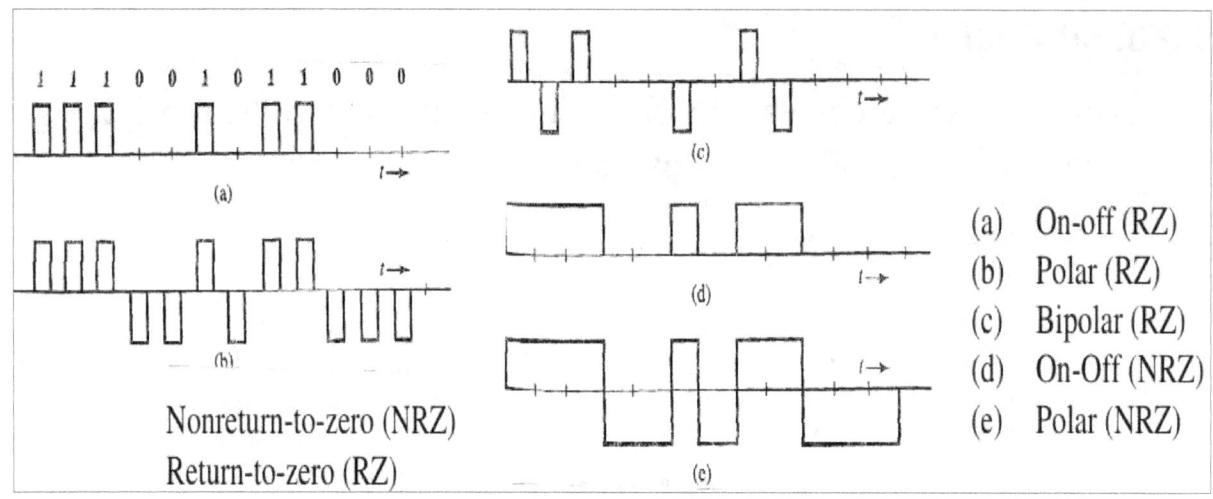

Pulse Transmission over a Channel

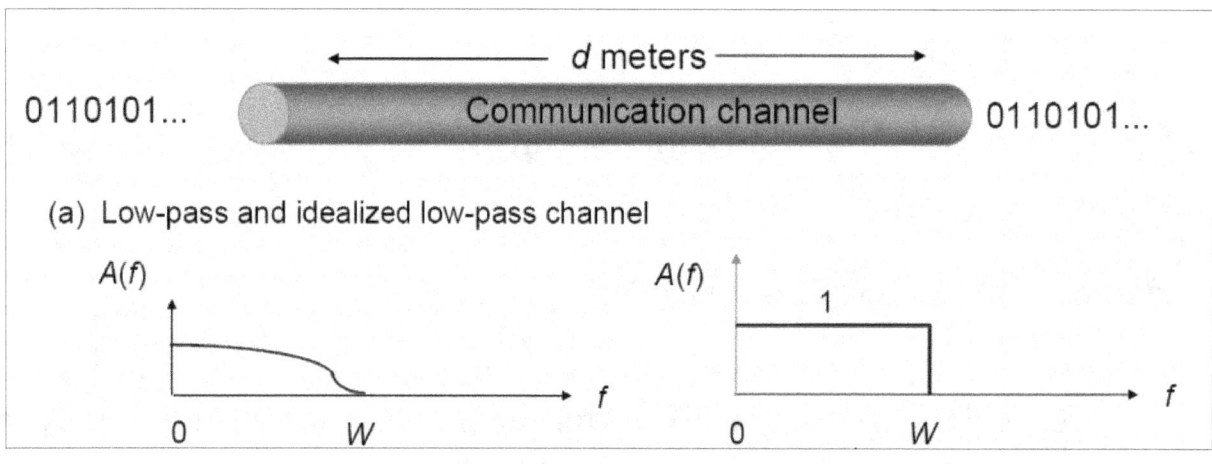

Desirable Properties for Line Codes

- Transmission Bandwidth: as small as possible
- Power Efficiency: As small as possible for given BW and probability of error
- Error Detection and Correction capability: Ex: Bipolar

- Favorable power spectral density: dc=0
- Adequate timing content: Extract timing from pulses
- Transparency: Prevent long strings of 0s or 1s

Review: Energy and Power Signals

- An energy signal x(t) has $0 < E < \infty$ for average energy

$$E = \int_{-\infty}^{\infty} |x(t)|^2 dt$$

- A power signal x(t) has $0 < P < \infty$ for average power

$$P = \lim_{T \to \infty} \frac{1}{2T} \int_{-T}^{T} |x(t)|^2 dt$$

- Can think of average power as average energy/time.
- An energy signal has zero average power.
- A power signal has infinite average energy.
- Power signals are generally not integrable so don't necessarily have a Fourier transform.
- We use power spectral density to characterize power signals that don't have a Fourier transform.

Review: Time-Invariant Systems

- Linear Time-Invariant Systems
- System Impulse Response: h(t)
- Filtering as Convolution in Time
- Frequency Response: $H(f) = |H(f)| e^{j < H(f)}$

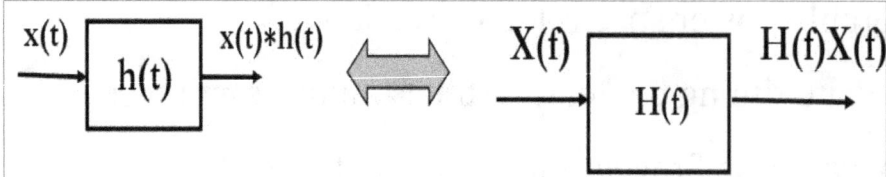

Review: Distortion

- Distortionless Transmission
    - Output equals input except for amplitude scaling and/or delay

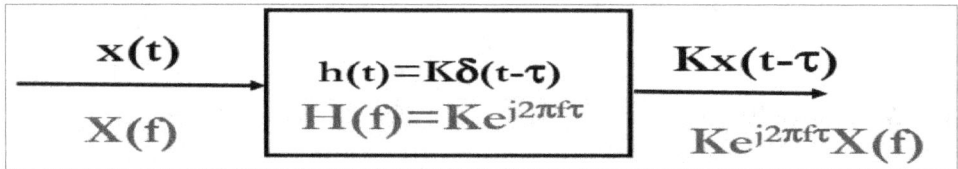

- Simple equalizers invert channel distortion
    - Can enhance noise power

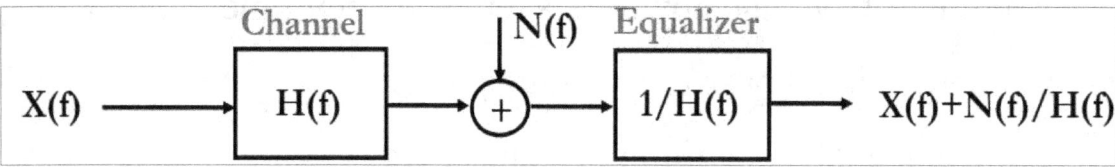

Review: Ideal Filters

- Low Pass Filter

- Band Pass Filter

Power Spectral Density

- Power signals (P=Energy/t)

- Distribution of signal power over frequency

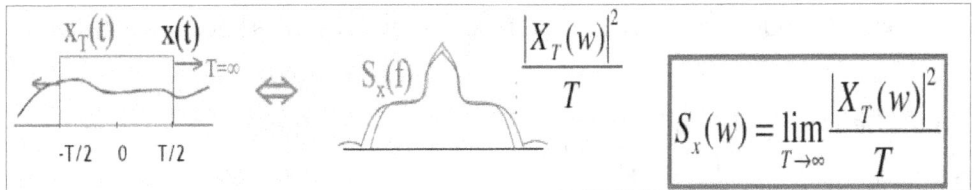

- Useful for filter analysis

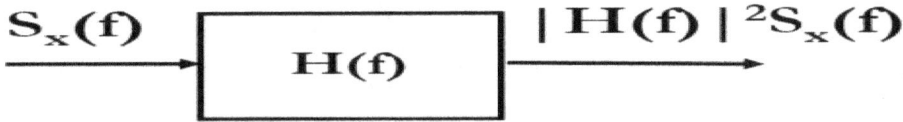

For Sx(f) bandlimited [−B,B], B<<fc

Definition: Autocorrelation

- Defined for real signals as $R_x(\tau) = x(\tau) * x(-\tau)$

$$R_x(\tau) = \lim_{T \to \infty} \frac{1}{T} \int_{-T/2}^{T/2} x(t) x(t-\tau) dt$$

- Measures similarity of a signal with itself as a function of delay

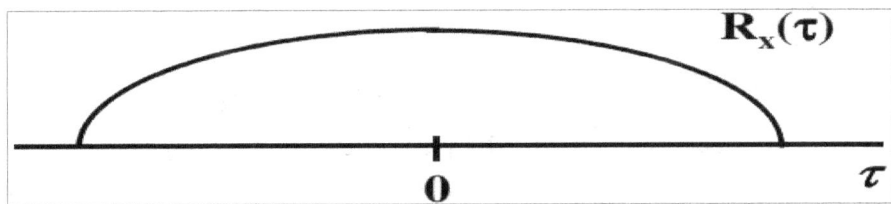

- Useful for synchronization: $|R_x(\tau)| \leq R_x(\tau)$
- PSD and autocorrelation FT pairs: $R_x(\tau) \Leftrightarrow S_x(f)$

Bandwidth Usage of Line Codes

- Line codes are used for digital base-band modulation in data communication applications,
- Digital data stream is encoded into a sequence of pulses for transmission through a base-band analog channel.

- The spectral properties of the line codes.
- We need a procedure for finding the PSD of line codes

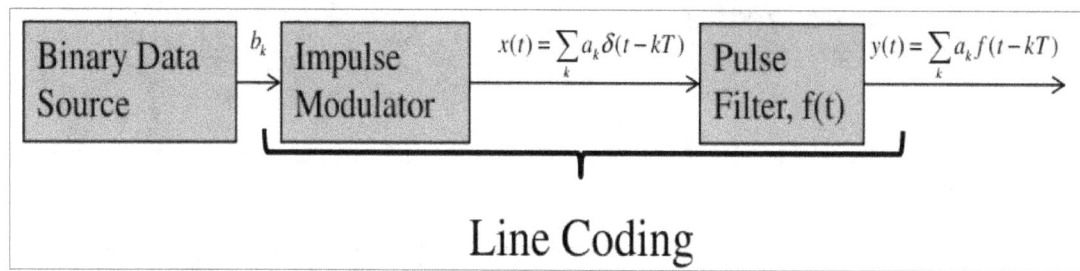

PSD Estimation

- We consider line coding pulses as a pulse train constructed from a basic pulse f(t) repeating at intervals of T with relative strength ak for the pulse starting at t=kT such that the kth pulse in this pulse train y(t) is akf(t-kT).

- For instance, the on-off, polar, and bipolar line codes are all special cases of this pulse train y(t), where a(k) takes on values 0,1, or -1 randomly subject to some constraints.

- We can analyze the various lines codes from the knowledge of the PSD of y(t)

- Simplify the PSD derivation by considering x(t) that uses a unit impulse response for the basic pulse of f(t)

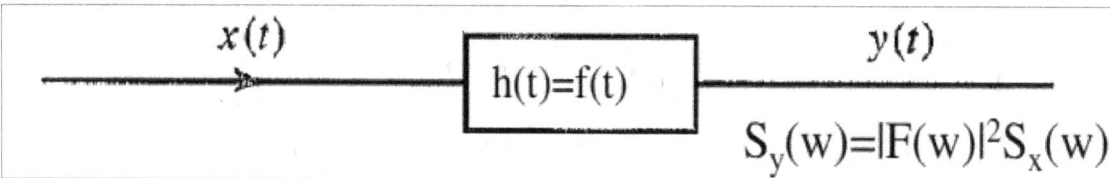

Power Spectral Density

- PSD is the Fourier Transform of autocorrelation
- Rectangular pulse and its spectrum

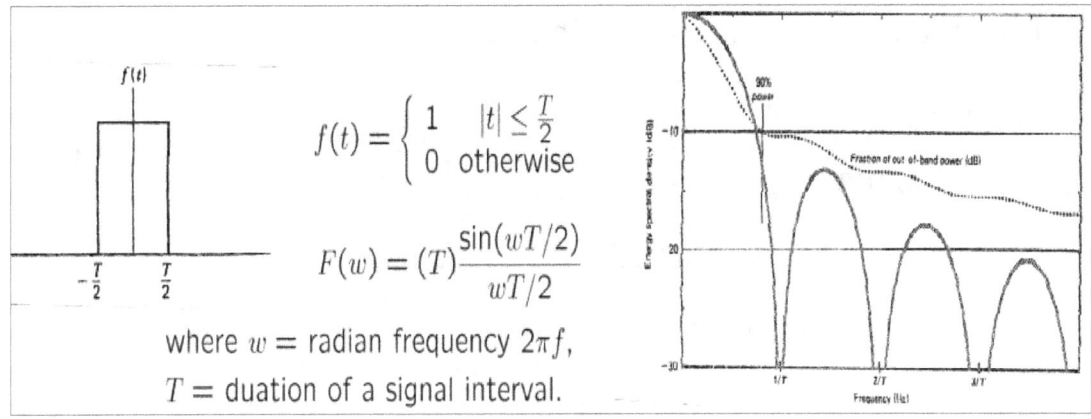

where w = radian frequency $2\pi f$,
T = duation of a signal interval.

PSD Derivation

- We now need to derive the time autocorrelation of a power signal x(t)

$$R_x(\tau) = \lim_{T_p \to \infty} \frac{1}{T_p} \int_{-T_p/2}^{T_p/2} x(t)x(t+\tau)dt$$

- Since x(t) consists of impulses, Rx(τ) is found by

$$R_x(\tau) = \frac{1}{T} \sum_{n=-\infty}^{\infty} R_n \delta(\tau - nT)$$

$$R_n = \lim_{N \to \infty} \frac{1}{N} \sum_k a_k a_{k+n}$$

- Recognizing Rn=R-n for real signals, we have

$$S_x(w) = \frac{1}{T}\left(R_0 + 2\sum_{n=1}^{\infty} R_n \cos nwT\right)$$

- Since the pulse filter has the spectrum of F(w) ↔ f(t), we have

$$\begin{aligned}
S_y(w) &= |F(w)|^2 S_x(w) \\
&= |F(w)|^2 \left(\sum_{n=-\infty}^{\infty} R_n e^{-jnwT_b}\right) \\
&= \frac{|F(w)|^2}{T}\left(R_0 + 2\sum_{n=1}^{\infty} R_n \cos nwT\right)
\end{aligned}$$

- Now, we can use this to find the PSD of various line codes.

PSD of Polar Signaling

- In polar signaling,
  - binary "1" is transmitted by a pulse f(t)
  - Binary "0" is transmitted by a pulse –f(t)
- In this case, a_k is equally likely to be 1 or -1 and a_k^2 is always 1

$$R_0 = \lim_{N \to \infty} \frac{1}{N} \sum_k a_k^2 = \lim_{N \to \infty} \frac{1}{N}(N) = 1$$

  - There are N pulses and $a_k^2=1$ for each one.
  - The summation on the right-hand side of the above equation is N.
- Moreover, both a_k and a_{k+1} are either 1 or -1. So, $a_k a_{k+1}$ is either 1 or -1.
- They are equally likely to be 1 or -1 on the average, out of N terms the product, $a_k a_{k+1}$ is equal to 1 for N/2 terms and is equal to -1 for the remaining N/2 terms.

$$R_1 = \lim_{N \to \infty} \frac{1}{N}\left[\frac{N}{2}(1) + \frac{N}{2}(-1)\right] = 0$$

$$R_n = 0 \qquad n \geq 1$$

$$S_y(w) = \frac{|F(w)|^2}{T} R_0 = \frac{|F(w)|^2}{T}$$

$$S_y(w) = \frac{T}{2}\text{sinc}^2\left(\frac{wT}{2}\right)$$

Bipolar Signaling

- Bipolar signaling is used in PCM these days.
- A "0" is transmitted by no pulse
- A "1" is transmitted by a pulse f(t) or –f(t), depending on whether the previous "1" was transmitted by –f(t) or f(t)

- With consecutive pulses alternating, we can avoid the dc wander and thus cause a dc null in the PSD. Bipolar signaling actually uses three symbols [f(t),0,-f(t)], and hence, it is in reality ternary rather than binary signaling.

- To calculate the PSD, we have

$$R_n = \lim_{N \to \infty} \frac{1}{N} \sum_k a_k a_{k+n} \qquad R_0 = \lim_{N \to \infty} \frac{1}{N} \sum_k a_k^2$$

- On the average, half of the a_k's are 0, and the remaining half are either 1 or -1, with $a_k^2=1$. Therefore,

$$R_0 = \lim_{N \to \infty} \frac{1}{N} \left[\frac{N}{2}(\pm 1)^2 + \frac{N}{2}(0)^2 \right] = \frac{1}{2}$$

- To compute R_1, we consider the pulse strength product $a_k a_{k+1}$.

- Four possible equally likely sequences of two bits: 11, 10, 01, 00.

- Since bit 0 encoded by no pulse ($a_k=0$), the product, $a_k a_{k+1}=0$ for the last three of these sequences. This means that, on the average, 3N/4 combinations have $a_k a_{k+1}=0$ and only N/4 combinations have non zero $a_k a_{k+1}$. Because of the bipolar rule, the bit sequence 11 can only be encoded by two consecutive pulse of opposite polarities. This means the product $a_k a_{k+1}= -1$ for the N/4 combinations

$$R_1 = \lim_{N \to \infty} \frac{1}{N} \left[\frac{N}{4}(-1) + \frac{N}{4}(0) \right] = -\frac{1}{4}$$

PSD of Lines Codes

- PSD of lines codes

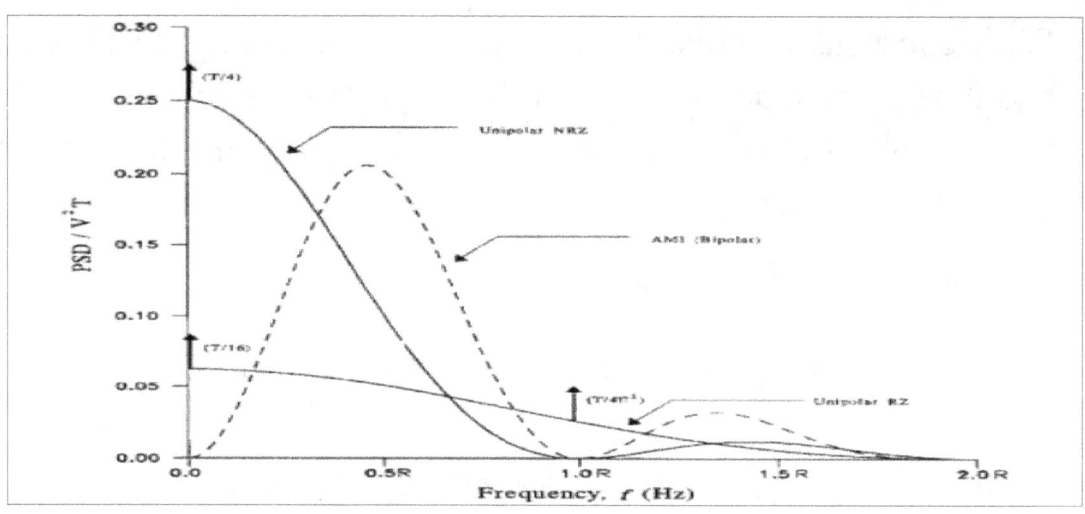

Binary N-zero Substitution (BNZS)

- Bipolar signaling has several advantages: (1) its spectrum has a dc null. (2) its bandwidth is not excessive. (3) it has single-error-detection capability. This is a due to the fact that if a single detection error is made, it will violate the alternating pulse rule.

- Disadvantages of bipolar signaling: it requires twice as much power (3 dB) as a polar signal. It is not transparent,
    - i.e., we need a minimum density of 1's in the source to maintain timing at the regenerative repeaters. Low density of pulses increases timing jitter.

- Solution: Binary N-zero substitution (BNZS) augments a basic bipolar code by replacing all things of N 0's with a special N-length code containing several pulses that purposely produce bipolar violations.

BNZS Line Codes

- High Density Bipolar (HDB) coding is an example of BNZS coding format. It is used in E1 primary digital signal

- HDB coding replaces strings of four 0's with sequences containing a bipolar violation in the last bit position. Since this coding format precludes strings of 0's greater than three, it is referred to as HDB3 coding.

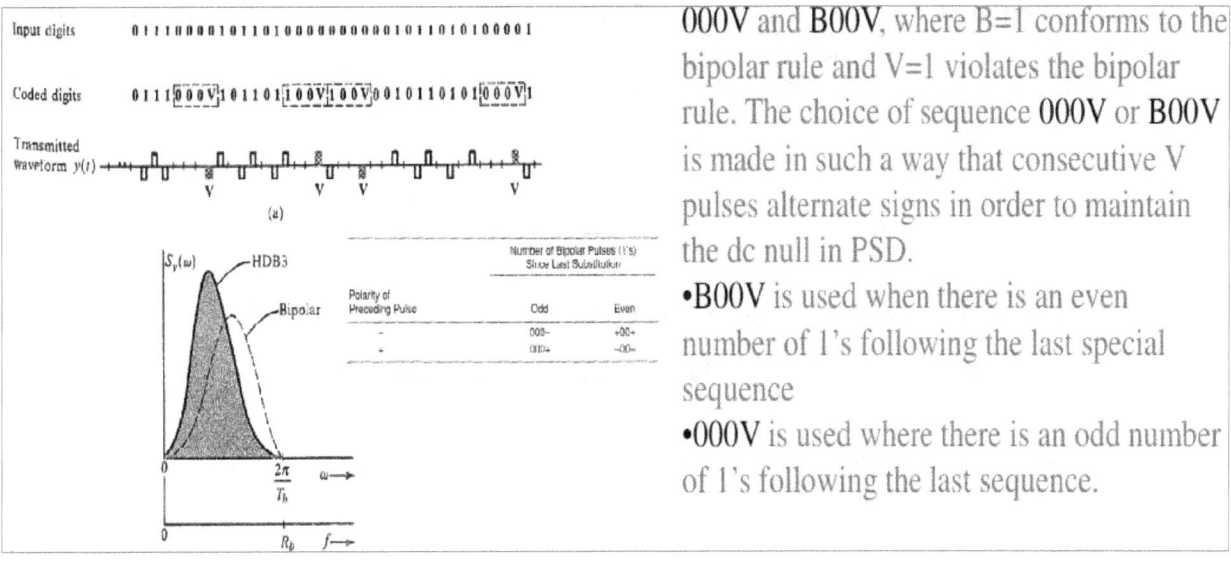

B3ZS Line Code

- B3ZS Algorithm (as used DS-3 signal interface): Each string of three 0's in the source data is encoded with either 00v or B0V.

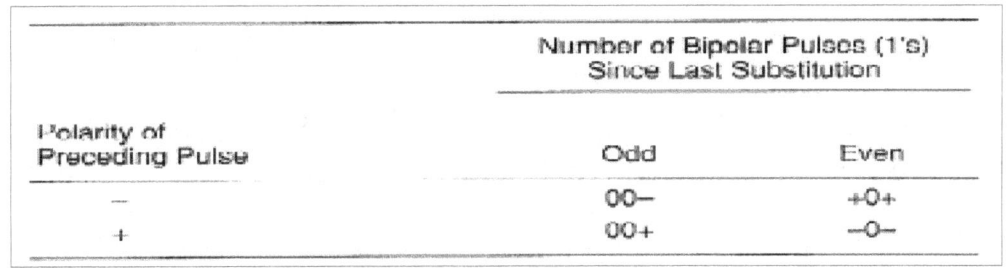

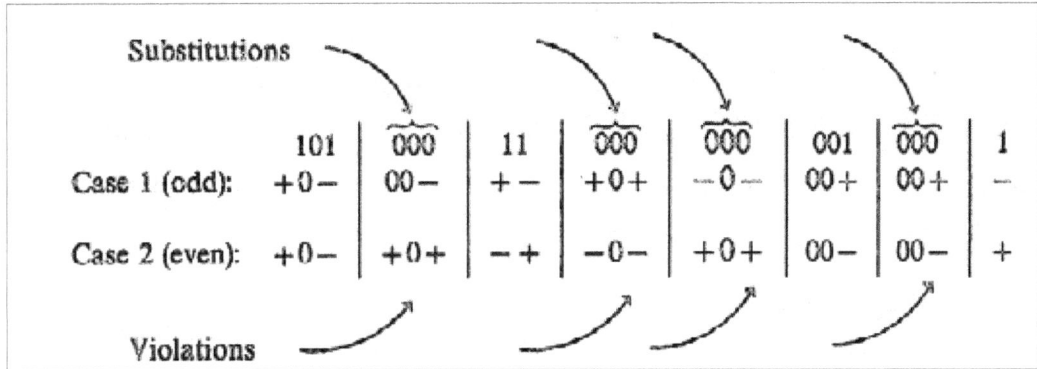

B8ZS Signaling

111

- B8ZS line code is used for T1 (DS1 signals) lines. It replaces any string of eight zeros in length with a sequence of 1's and 0's containing two bipolar violations. There are two bipolar violations in every substitution.

Polarity of Preceding Pulse	Substitution
−	000−+0+−
+	000+−0−+

- Similarly, in B6ZS code used in DS2 signals, a string of six zeros is replaced with 0VB0VB.

Polarity of Pulse Immediately Preceding Six 0's to be Substituted	Substitution
−	0−+0+−
+	0+−0−+

Example:
```
  1 0 0 0 0 0 0 1 0 1 1 0 0 0 0 0 0 0 0 0 0 0 0 0 0 1
+ − (0 − + 0 + −) + 0 − + (0 + − 0 − +) (0 + − 0 − +) 0 0 0 −
− + (0 + − 0 − +) − 0 + − (0 − + 0 + −) (0 − + 0 + −) 0 0 0 +
```

Differential Encoding:

- One limitation of polar signaling is that the signal for a 1 is exactly the negative of a signal for a 0. On many transmissions, it may be impossible to determine the exact polarity or an absolute phase reference.

- The decoder may decode all 1's as 0's or vice versa.

- Common remedy for the phase ambiguity is to use differential encoding that encodes a 1 as a chance of states and encodes a 0 as no change in state. In this way, we do not need absolute phase reference.

Differential Encoding

- The differentially encoded sequence {d$_k$} is generated from the input binary sequence {m$_k$} by complementing the modulo-2 sum of m$_k$ and d$_{k-1}$. The effect is leaving the symbol d$_k$ unchanged from the previous symbol if the incoming binary symbol m$_k$ is 1, and to toggle d$_k$ if m$_k$ is 0.

$$d_k = \overline{m_k \oplus d_{k-1}}$$

{m$_k$}		1	0	0	1	0	1	1	0
{d$_{k-1}$}		1	1	0	1	1	0	0	0
{d$_k$}	1	1	0	1	1	0	0	0	1

- the decoder merely detects the state of each signal interval and compares it to the state of the previous signal.
  - ➢ If changed occurred, a 1 is decoded. Otherwise, a 0 is determined.

$$m_k = \overline{d_k} \oplus d_{k-1}$$

d$_k$	1	1	0	1	1	0	0	0	1
$\overline{d_k}$		0	0	1	0	0	1	1	1	0
d$_{k-1}$			1	1	0	1	1	0	0	0
m$_k$			1	0	0	1	0	1	1	0

Applications of Line Coding

- NRZ encoding: RS232 based protocols
- Manchester encoding: Ethernet networks
- Differential Manchester encoding: token-ring networks
- NRZ-Inverted encoding: Fiber Distributed Data Interface (FDDI)

Asynchronous vs Synchronous Transmission

- Asynchronous transmission: Separate transmissions of groups of bits or characters
- The sample clock is reestablished for each reception
- Between transmissions an asynchronous line is in idle state.

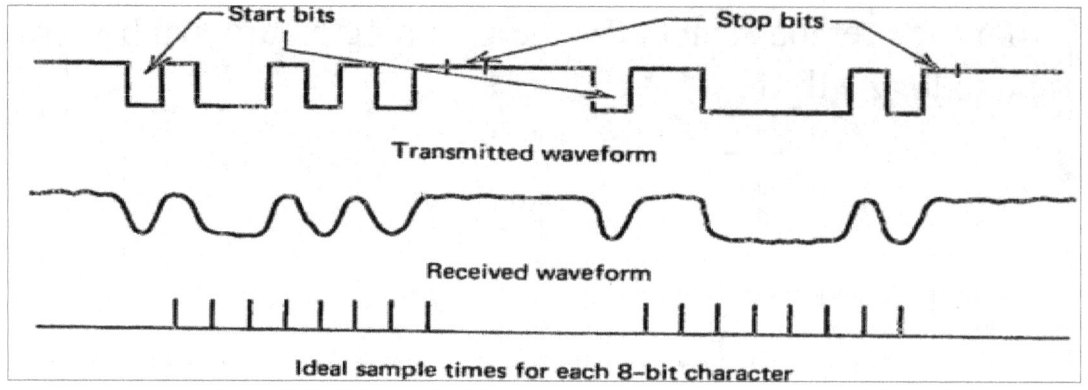

- Synchronous transmission: Digital signals are sent continuously at a constant rate

  ➢ The sample clock is established and maintained throughout entire time.

- Synchronization Consideration

  ➢ Problem of unvarying signal:

  - When a signal is unvarying, the receiver cannot determine the beginning and ending of each bit.

  - Take unipolar coding for example. A long uninterrupted series of 1s or 0s can cause synchronization problem.

  ➢ Problem of Using Timers: Whenever there is no signal change to indicate the start of the next bit in a sequence, the receiver has to rely on a timer. Given an expected bit rate of 1000 bps, if the receiver detects a positive voltage lasting 0.005 seconds, it reads one 1 per 0.001 seconds, or five 1s. However, five 1s can be stretched to 0.006 second, causing an extra 1 bits to be read by the receiver. That one extra bit in the data stream causes everything after it to be decoded erroneously.

  ➢ Problem of Having a Separate Clock Line: A solution developed to control the synchronization of unipolar transmission is to use a separate, parallel line that carries a clock pulse. But doubling the number of lines used for transmission increase the cost.

Synchronous Communication

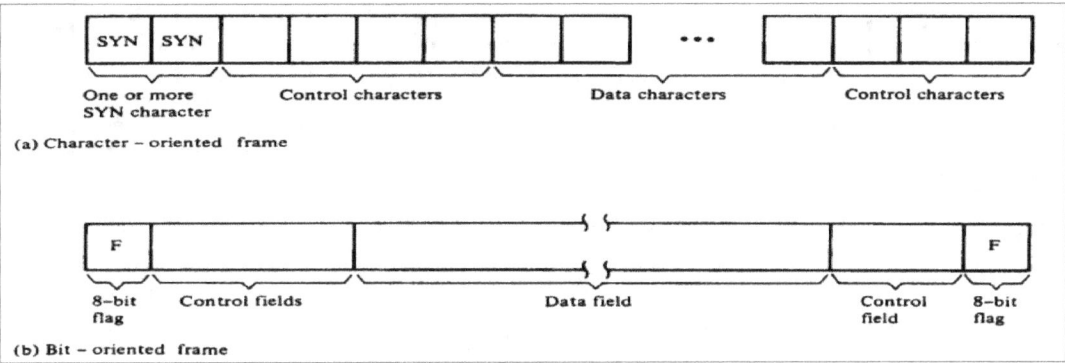

(a) Character – oriented frame

(b) Bit – oriented frame

Asynchronous Transmission

- Bits are sent one character at a time. (A character is in general 8 bits in length)

- Timing or synchronization must only be maintained within each character. The receiver has the opportunity to resynchronize at the beginning of each new character.

Start-stop technique

- Idle state: When no character is being transmitted the line between transmitter and receiver is in an "idle" state. The definition of idle is by convention, but typically is equivalent to the signaling element for binary 1.

Asynchronous Transmission

- Start bit: The beginning of a character is signaled by a start bit with a value of binary 0.

- Data bits

- Stop bit: The last bit of the character is followed by a stop bit, which is a binary 1. A minimum length for the stop bit is specified and this is usually 1, 1.5 or 2 times the duration of an ordinary bit. No maximum value is specified, Since the stop bit is the same as the idle state.

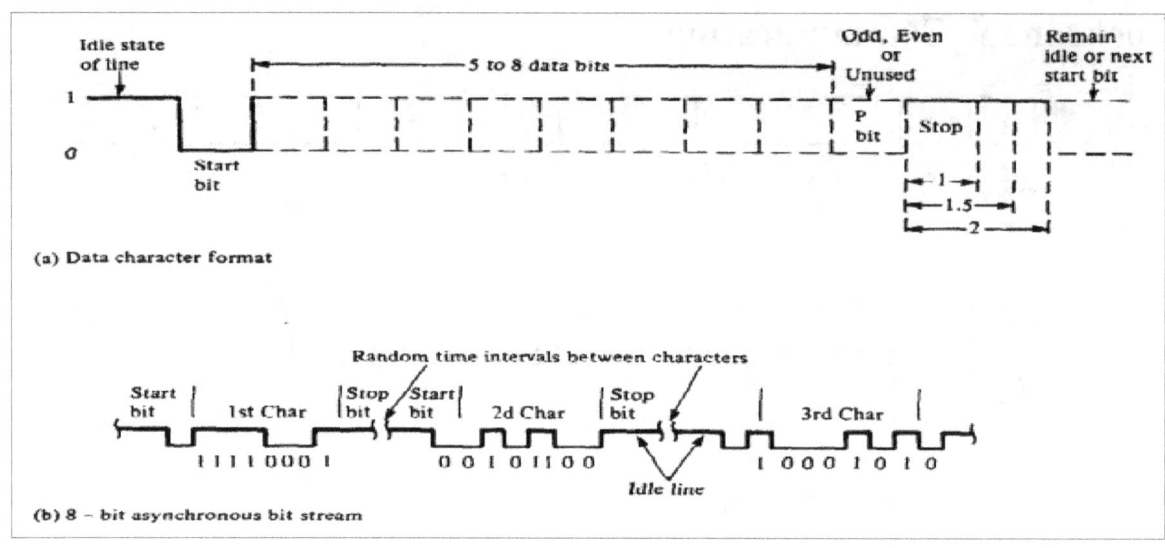

(a) Data character format

(b) 8 - bit asynchronous bit stream

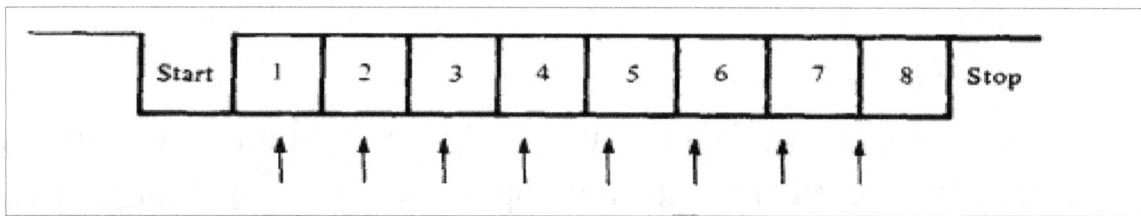

Bandwidth Definitions:

Measures of Bandwidth (BW):

- 99% BW freq. range where 99% of power is
- Absolute BW: Range of frequencies over a non-zero spectrum
- Null-to-Null BW: Width of the main spectral lobe
- Half-power bandwidth: 3dB bandwidth

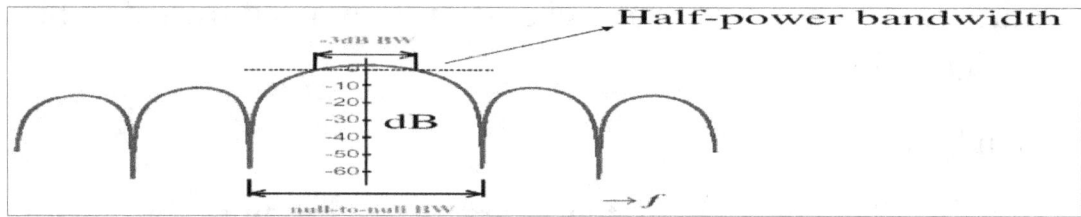

Pulse Shaping:

- Pulse shaping concerns with how to shape a pulse p(t) in order to achieve a desired $S_y(w)$.

- The PSD $S_y(w)$ is strongly and directly influenced by the pulse shape f(t) because $S_y(w)$ contains the term $|F(w)|^2$.
- Typical pulse response of a bandlimited channel

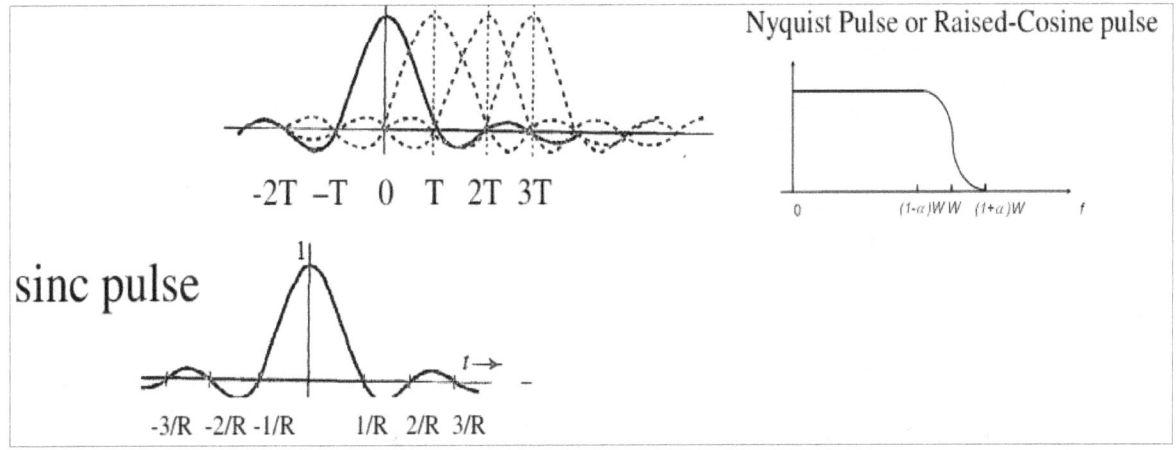

Maximum Signaling Rate

- The percentage of total spectrum power is important measure

A major result for digital transmission pertains to the maximum rate at which pulses can be transmitted over a channel.

- If a channel has bandwidth W, then the narrowest pulse that can be transmitted over the channel has duration T=1/(2W) seconds.

Thus, the maximum rate at which pulses can be transmitted through the channel is given by R_{max}=2×W pulses/second

Multilevel Signaling

- Digital communications use only a finite number of symbols for communication, the minimum being two (binary)
- Thus far, we have only considered the binary case. In some applications, the bandwidth is limited but higher data rates are desired, number of symbols (i.e., voltage levels) can be increased while maintaining the same signaling rate (baud rate).

- Multilevel signaling: The data rate R achieved by a multilevel system is given by

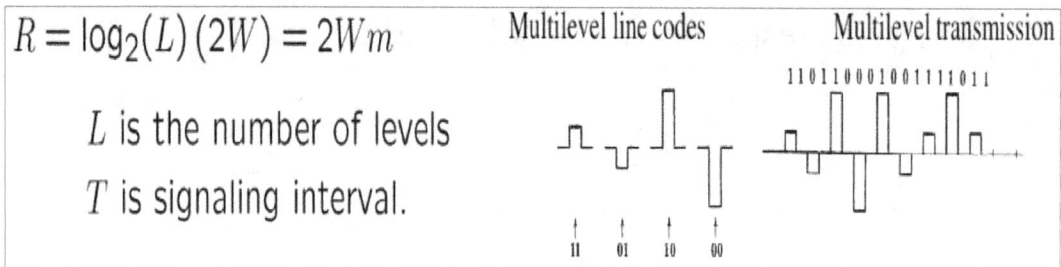

$R = \log_2(L)(2W) = 2Wm$

L is the number of levels

T is signaling interval.

Multilevel Signaling and Channel Capacity

- Suppose we increase the number of levels while keeping the maximum signal levels ±A fixed. Each increase in the number of signal levels requires a reduction in the spacing between levels. At some point, these reductions will imply significant increases in the probability of detection errors as the noise will be more likely to cause detection errors

- The channel capacity of a transmission system is the maximum rate at which bits can be transferred reliably. Shannon derived an expression for channel capacity of an ideal low-pass channel. He showed that reliable communication is not possible at rates above this capacity.

$$C = W \log_2(1 + \text{SNR}) \quad \text{bits/second}$$

Multi-Level Signals and Noise:

Typical noise

Four signal levels   Eight signal levels

Signal-toto-Noise Ratio

- Definition of SNR

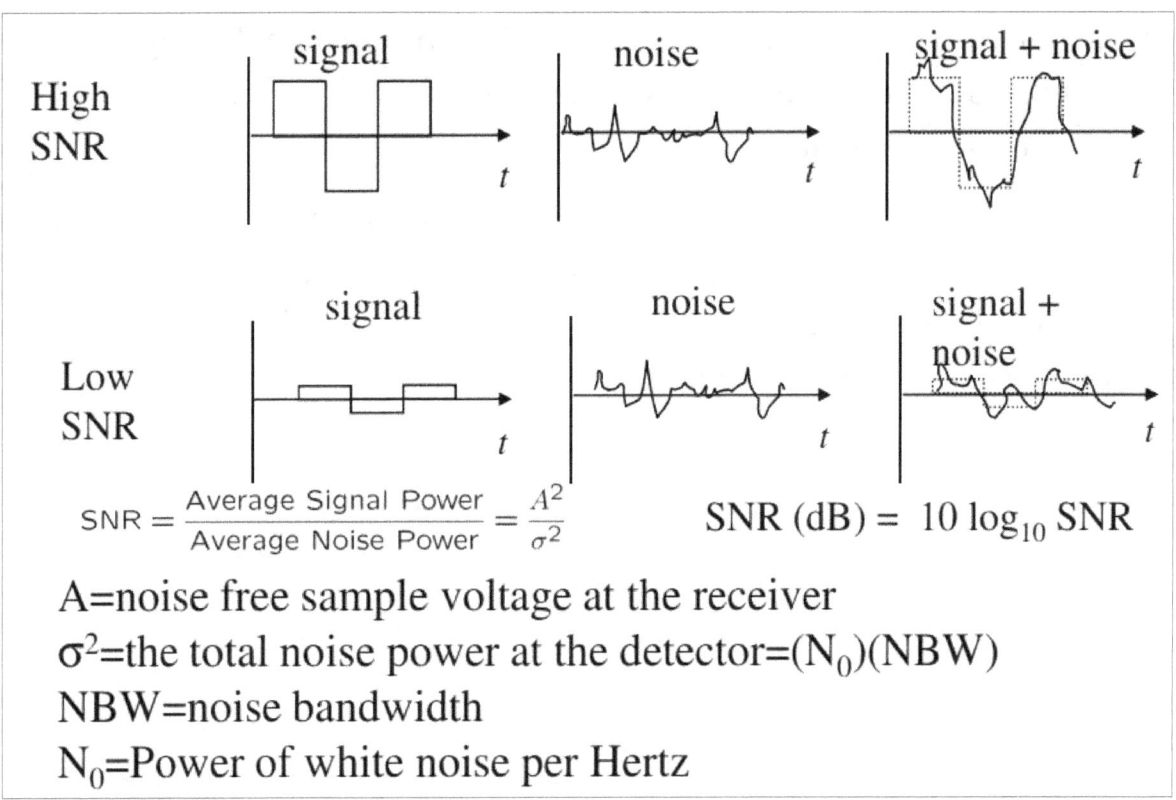

$$\text{SNR} = \frac{\text{Average Signal Power}}{\text{Average Noise Power}} = \frac{A^2}{\sigma^2} \qquad \text{SNR (dB)} = 10 \log_{10} \text{SNR}$$

A = noise free sample voltage at the receiver
σ^2 = the total noise power at the detector = $(N_0)(NBW)$
NBW = noise bandwidth
N_0 = Power of white noise per Hertz

Error Performance

- Signal Detection: A decision of which signal was transmitted is made by comparing the measurement (at the appropriate time) to a threshold located halfway between these nominal voltages that represent "0" and "1".

- Error performance depends on the nominal distance between the voltages and the amount of fluctuation in the measurements caused by noise.

- In absence of noise, the measurement of the positive pulse would be A and that of negative pulse would be –A. Because of noise, these samples would be $\mp A + n$ where n is the random noise amplitude.

- The error performance analysis in communication circuits is typically based on white Gaussian noise.

Error Probabilities

- We now compute the probability of error for a polar signal. The amplitude n of the noise is Gaussian distributed. It ranges from -∞ to ∞ according Gaussian PDF.

- When "0" is transmitted, the sample value of the received pulse is −A+n. If n>A, the sample value is positive and the digit will be detected wrongly as 1. If P(error|0) is the probability of error given that 0 is transmitted, then,

$$P(\text{error}|0) = Prob(n < -A) = \frac{1}{\sigma\sqrt{2\pi}} \int_A^\infty e^{-n^2/2\sigma^2} dn \qquad Q(y) = \frac{1}{\sqrt{2\pi}} \int_y^\infty e^{-x^2/2} dx$$

$$= \frac{1}{\sqrt{2\pi}} \int_{A/\sigma}^\infty e^{-x^2/2} dx$$

$$= Q\left(\frac{A}{\sigma}\right)$$

- Probability of error for a polar signal

$$P(\text{error}) = \frac{1}{2}[P(\text{error}|0) + P(\text{error}|1)] = Q\left(\frac{A}{\sigma}\right)$$

Twisted Pair

- A twisted pair consists of two wires that are twisted together to reduce the susceptibility to interference.

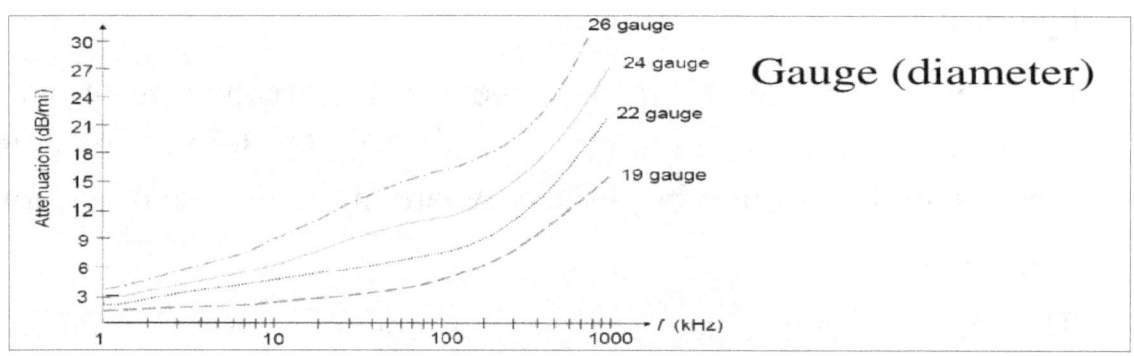

- The two-wire system is susceptible to crosstalk and noise since the multiple wires are bundled together

Polar Signaling

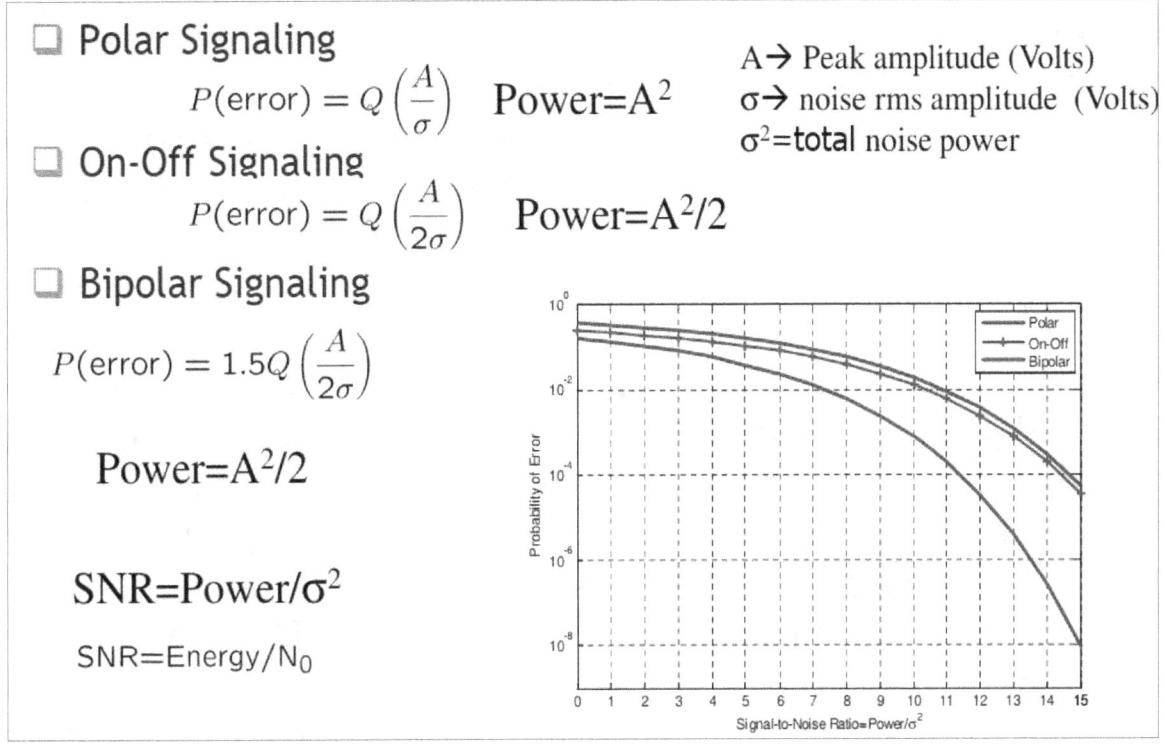

❏ Polar Signaling
$$P(\text{error}) = Q\left(\frac{A}{\sigma}\right) \quad \text{Power} = A^2$$

❏ On-Off Signaling
$$P(\text{error}) = Q\left(\frac{A}{2\sigma}\right) \quad \text{Power} = A^2/2$$

❏ Bipolar Signaling
$$P(\text{error}) = 1.5 Q\left(\frac{A}{2\sigma}\right)$$

Power=$A^2/2$

SNR=Power/σ^2

SNR=Energy/N_0

$A \rightarrow$ Peak amplitude (Volts)
$\sigma \rightarrow$ noise rms amplitude (Volts)
σ^2 = **total** noise power

Redundancy Checks

- Parity Bits are inserted into DS3 and DS4 signals for the purpose of monitoring the channel error rate.

- The following equation relates the parity error rate (PER) to the channel probability of error or bit error rate (BER)

$$\text{PER} = \sum_{i=1}^{N} \binom{N}{i} p^i (1-p)^{N-i} \quad (i \text{ odd})$$

N=number of bits over which parity is generated
p=BER assuming independent errors

- Cyclic redundancy check (CRC) codes are also incorporated into a number of transmission systems as a means of monitoring BERs and validating framing acquisition.

- Examples of CRC use: Extended superframe (ESF) on T1 lines

$$\text{CRCER} = 1 - (1-p)^N$$

N=length of CRC field (including CRC bits)
p=BER assuming independent errors

TDM and Codecs

Time Division Multiplexing (TDM) is the time interleaving of samples from several sources so that the information from these sources can be transmitted serially over a single communication channel.

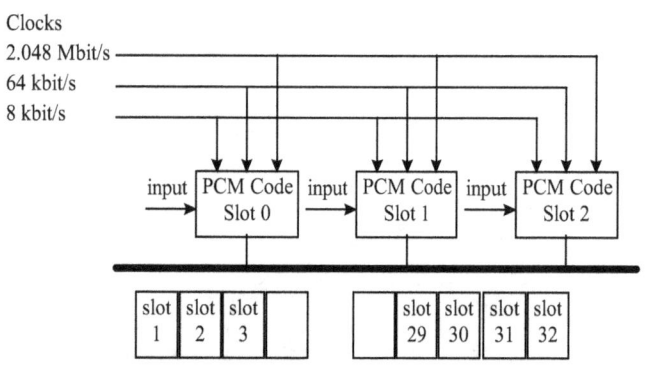

The diagram illustrates the TDM concept applied to 3 PCM **Codecs**. In this system the PCM **cod**er samples the analogue signal at 8 kHz and generates an 8-bit code for each sample. The coder maps the 8-bit code onto a 2,048 Mbit/s bus at a precise time. This time is determined by the slot that the coder is programmed to operate on. The process is repeated every 125 μs. Obviously the codecs must be exactly synchronised to the master PCM bus. The PCM **dec**oder simply reads the data in the particular time slot of interest.

This method of multiplexing is popular in telephone systems and there are many chip sets available to perform the above functions.

Lecture 3
Switching

3.1 Introduction

- A switch transfers signals from one input port to an appropriate output.

- A basic problem is then how to transfer traffic to the correct output port.

- In the early telephone network, operators closed circuits manually. In modern circuit switches this is done electronically in digital switches.

- If no circuit is available when a call is made, it will be blocked (rejected). When a call is finished a connection, teardown is required to make the circuit available for another user.

3.2 Crossbar Switch

- A crossbar switch with N input lines and N output lines contains an N x N array of cross points that connect each input line to one output line. In modern switches, each cross point is a semiconductor gate.

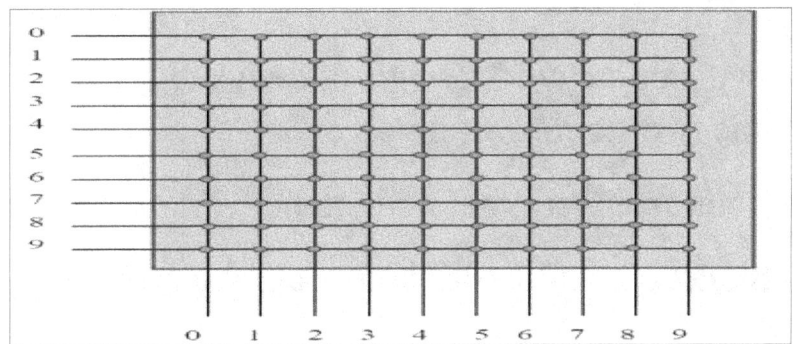

Switching Functions:

- Recall basic elements of communications network:
- Terminals, transmission media, and switches
- Basic function of any switch is to set up and release connections between transmission channels on an "as needed basis"
- Computers are used to control the switching functions of a central office

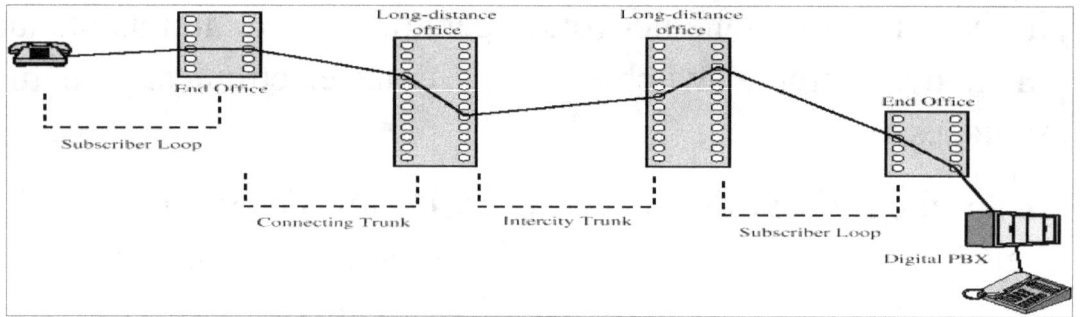

Switching Types

Two different switching technologies:

- Circuit switching
- Packet switching

Circuit-Switched Network

- Circuit-Switched network assigns a dedicated communication path between the two stations. It involves
- Point to Point from terminal node to network
- Internal Switching and multiplexing among switching nodes.
- Data Transfer.
- Circuit Disconnect.
- Advantages

Once connection is established

> Network is transparent.

> Nodes seems to be directly connected.

> Fixed data rate with no delay.

- Disadvantages

> Can be inefficient

> Resources are dedicated to

- Connection even if no data is sent.
- Delay prior to usage of connection

3.3 Space Division Switching

- Developed for analog environment
- Separate physical paths
- Recall Cross bar switch
- The no. of cross points grows with square of the lines
- attached. N×N array of cross points
- The loss of cross point means the loss of connection between the corresponding points.
- Only fraction of the cross points are used even when all the points are fully active. (sqrt of cross points)
- Non-blocking switching type.
- Less signaling requirement from the network.

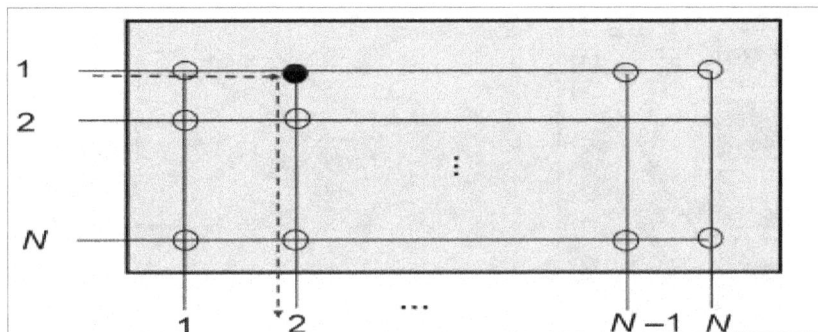

Multistage Switches

- Multistage switch
- Less no. of cross points are needed.
- More than one route for a connection.
- More signaling from the network.

- A blocking switching type (voice)

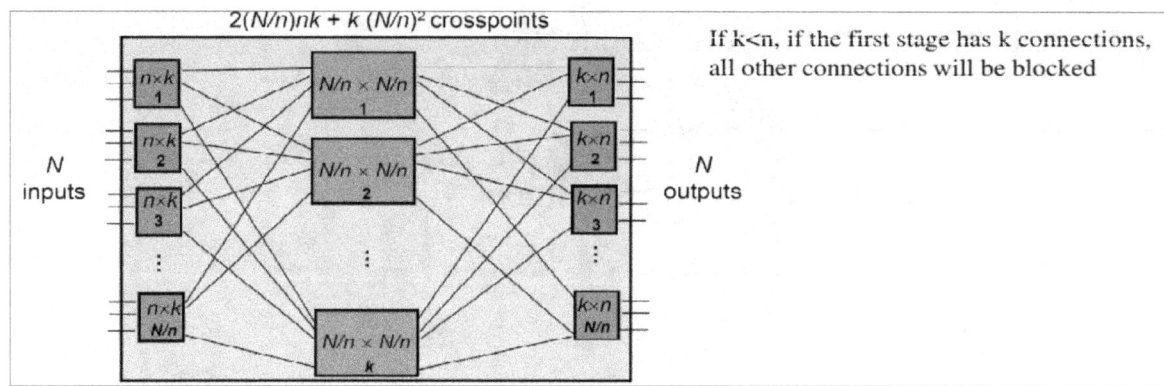

3.4 Nonblocking Switching:

- When a multistage switch becomes nonblocking?
- The multistage switch with k=2n-1 is nonblocking
- the number of cross points required in a three-stage switch is the sum of the following components

$$N/n \times nk + k \times (N/n)^2 + N/n \times nk = 2Nk + k(N/n)^2$$

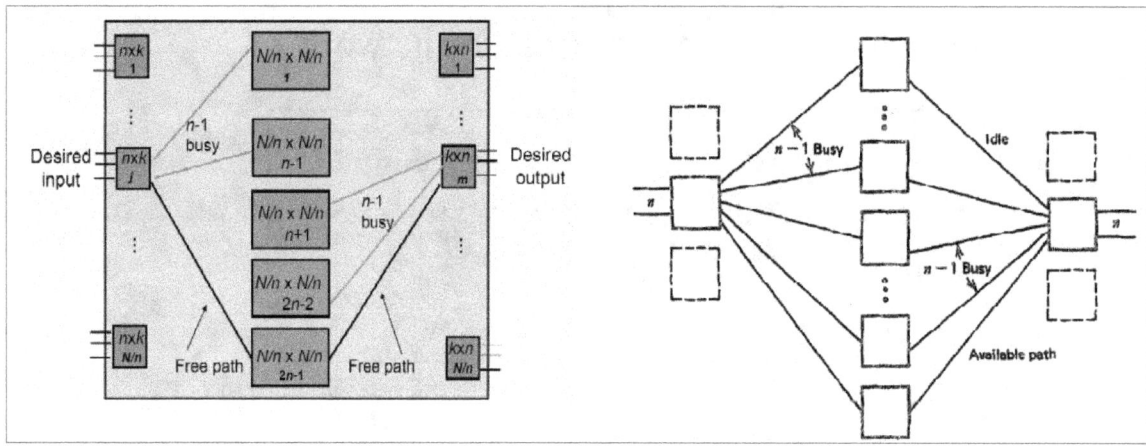

Blocking Probabilities:

- Strictly nonblocking switches are rarely needed in most voice telephone networks.
- Switching systems and the number of circuits in interoffice trunk groups are sized to service most requests (not all) as they occur

Economics dictates that network implementations have limited capacities that occasionally exceeded during peak-traffic situations Equipment for the public telephone network is designed to provide a certain maximum probability of blocking for the busiest hour of the day.

- Grade of service of the telephone company depends on the blocking probability, availability, transmission quality, and delay
- Residential lines are busy 5-10% of the time during the busy hour
- Network-blocking occurrences on the order of 1% during the busy hour do not represent a significant reduction in the ability to communicate since the called party is much more likely to have been busy anyway.

Evaluation of Blocking Probability:

- Probability graphs as proposed C. Y. Lee
  - Simplifying approximations are needed
  - Formulas directly relate to the underlying network structures
- Notation
  - p represents the fraction of the time that a particular link is in use (or p is the probability that a link is busy)
  - q=1-p is the probability that the link is idle.
  - When any one of n parallel links can be used to complete a connection, the composite blocking probability B is the probability that all links are busy

$$B = p^n$$

  - When a series of n links are all needed to complete a connection, the blocking probability is mostly determined as 1 minus the probability that they are all available

$$B = 1 - q^n$$

Probability Graph

- Any particular connection can be established with k different paths
  - One through each center-stage array

> B = probability that all paths are busy
> = probability that an arbitrary path is busy
> = probability that at least one link in a path is busy)^k
> = $(1 - (q')^2)^k$
>   where k=number of center-stage arrays
>   q'=probability that an interstage link is idle, =1-p'

- If the probability p that an inlet is busy is known, the probability p' that an interstage link is busy can be determined as

$$p' = \frac{p}{\beta} \quad (p < \beta) \quad \text{where} \quad \beta = k/n$$

- There are β=k/n times as many interstage links as there are inlets and outlets. The percentage of interstage links that are busy is reduced by the factor β. If β is less than 1, then the first stage is concentrating the incoming traffic.

3.5 Three-Stage Switch Design

- The blocking probability of a three-stage switch in terms of the inlet utilization p:

$$B = \left[1 - \left(1 - \frac{p}{\beta}\right)^2\right]^k$$

Time-Division Switching

- Mostly all modern circuit switches are time-division switches.
- Time-slot interexchange (TSI)
- It is based on synchronous TDM.

- Multiple low speed inputs share a high-speed line.
- There is no need for address bits in each slot (synchronous)
- The slot could be a bit, a byte or a longer block.

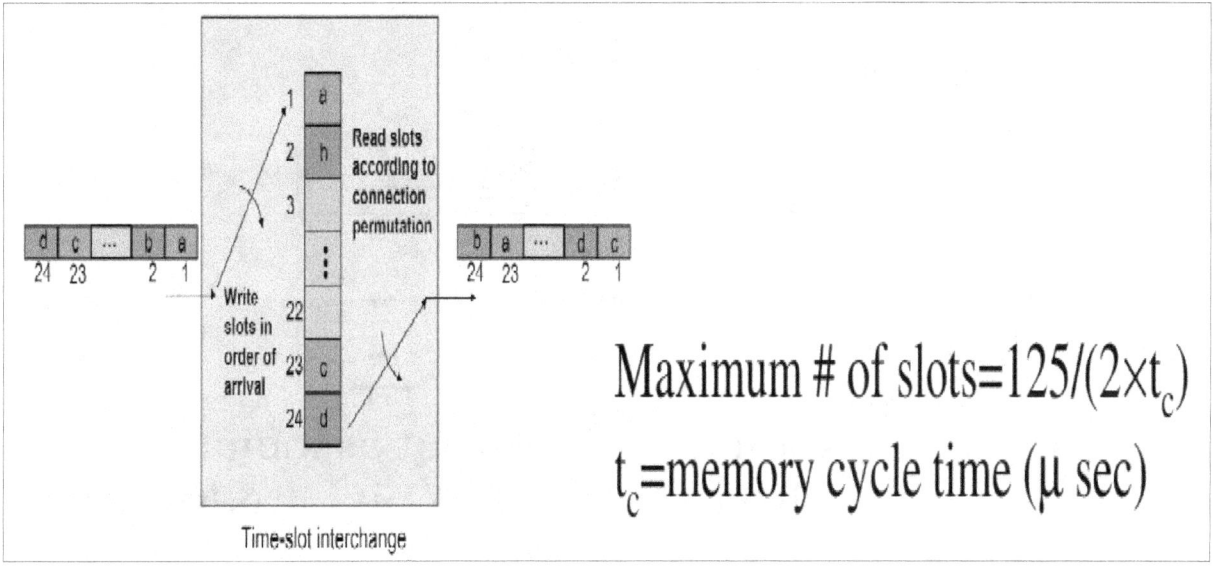

Maximum # of slots = $125/(2 \times t_c)$
t_c = memory cycle time (μ sec)

Time-slot interchange

MUX/TSI/DEMUX

- Incoming data slots are written into sequential locations of the data store memory.
- Data words from outgoing time slots, are read from addresses obtained from a control store

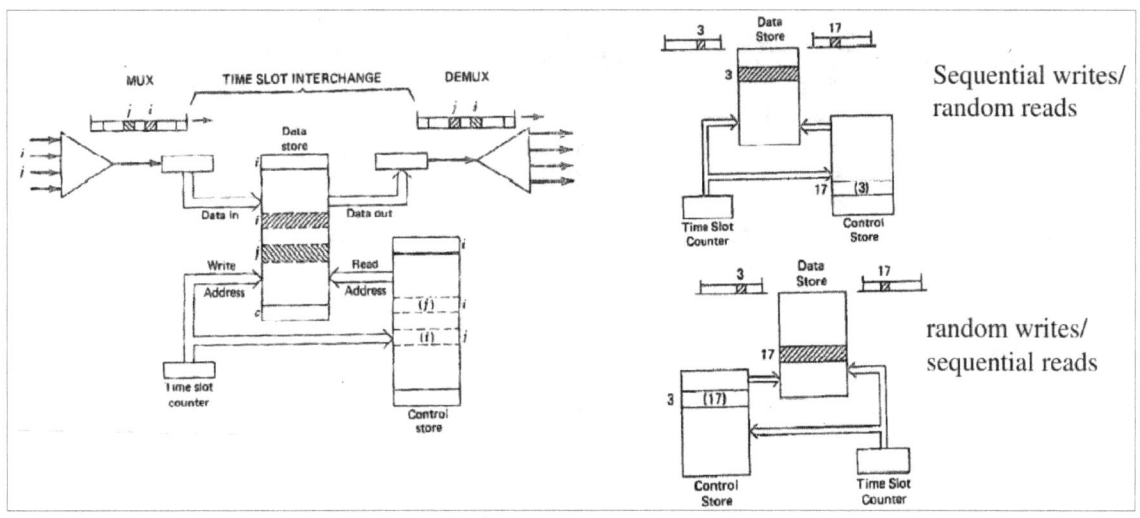

Switch Matrix Control

- Crosspoint selection within a matrix is accomplished in one of two ways.
- Input-associated control
- Output-associated control

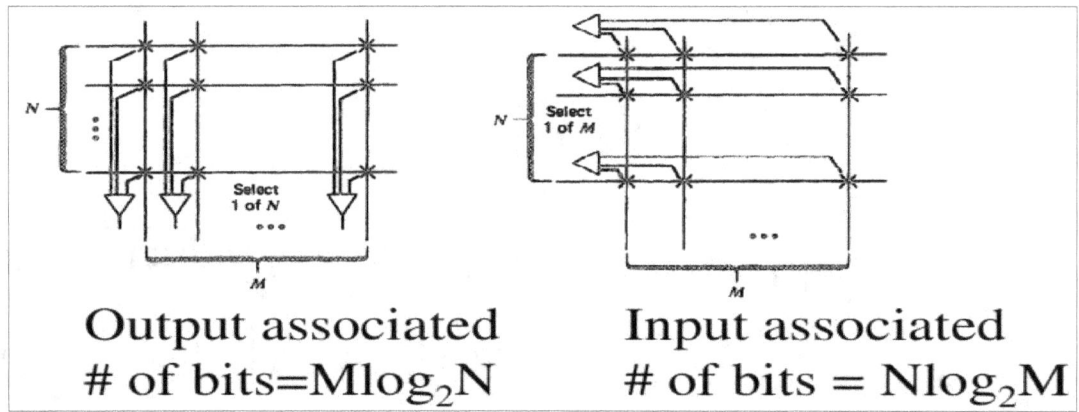

Output associated
# of bits=$M\log_2 N$

Input associated
# of bits = $N\log_2 M$

Hybrid Switches

- Hybrid switch design (or two-dimensional switching)
- Time-Space switch
- Space-Time-Space switch
- Time-Space-Time switch

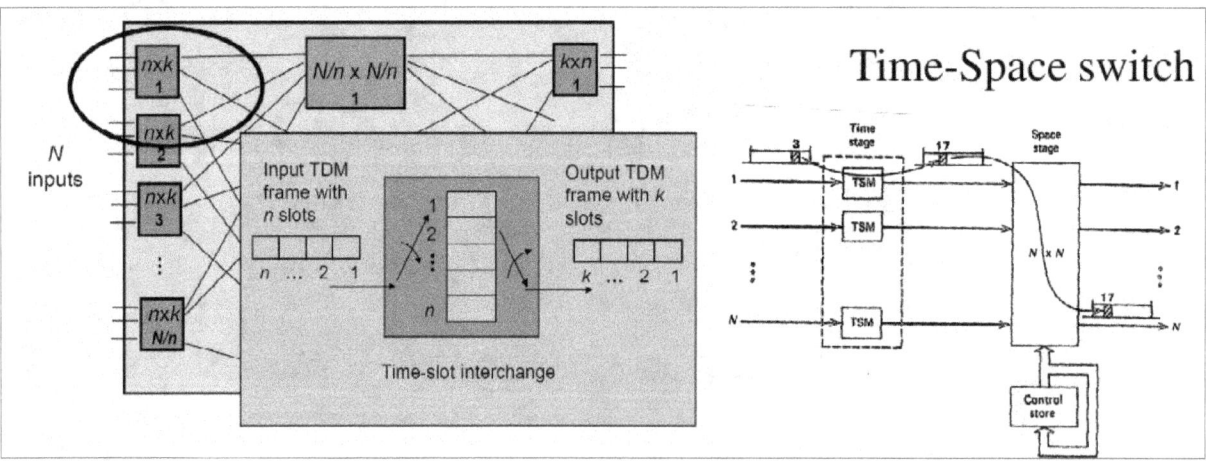

3.6 Implementation Complexity of TDS

- Total number of crosspoints alone is a less meaningful measure of implementation cost
- We have to include cost of the implementation including control bits
- Cost of number bits vs cost of crosspoints, (we use the ratio as 100)
- Complexity = $N_X + N_B/100$
- N_X = Number of space stage crosspoints
- N_B is number of bits of memory and control

Implementation Complexity Example

- Determine the implementation complexity of the TS switch shown in previous slide:
- # of TDM input lines N=80
- Each input contains a single DS1 signal (24 channels).
- Assume a one-stage matrix is used for the space stage
- Number of cross points: $N_X = 8^2 = 6400$

Implementation Complexity

- Total number of memory bits
- space stage control store N_{BX} = (number of links) (number of control words) (number of bits per control word)

$$N_{BX} = (80)(24)(7) = 13{,}440$$

- Time stage N_{BT} = time slot interchange memory + control = (number of links) *number of channels) (number of bits per channel) + (number of links) (number of control words) (number of bits per control word)

$$N_{BT} = (80)(24)(8)+(80)(24)(5)=24960$$
Complexity=N_X+(N_{BX}+N_{BT})/100=6784 equivalent crosspoints

Space-Time-Space Switch

- Blocking probability of an STS switch

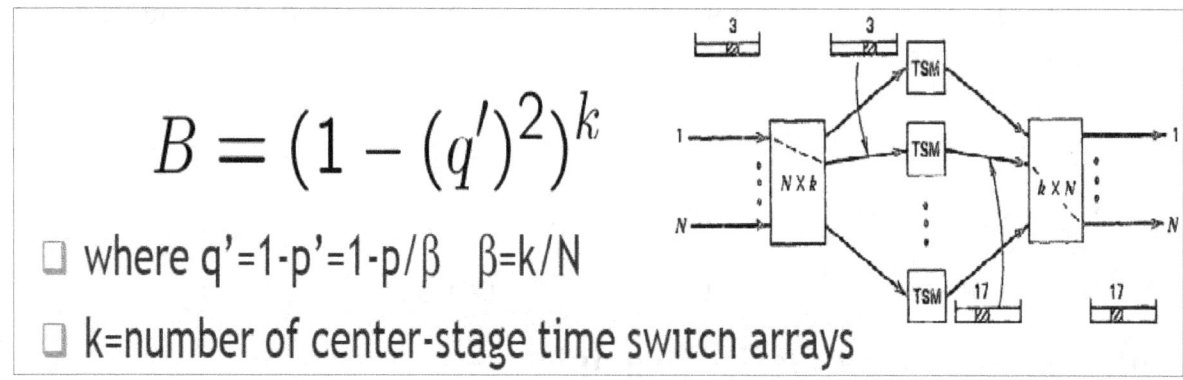

$$B = (1-(q')^2)^k$$

- where q'=1-p'=1-p/β   β=k/N
- k=number of center-stage time switch arrays

- Assume that each TDM link has c message channels Complexity of STS switch= number of space stage crosspoints + (number of space stage control bits + number of time stage memory bits+ number of time stage control bits)/100

Complexity=2kN+(2kc$\log_2$N+kc(8)+kc$\log_2$ c)/100

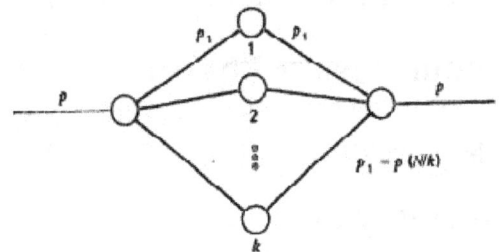

Example:

- Determine the implementation complexity of a 2048- channel STS switch implemented for 16 TDM links with 128 channels on each link. The desired maximum blocking probability is 0.002 for channel occupancies of 0.1

- k=7, B=0.002
- $N_X=(2)(7)(16)=224$
- $N_B=(2)(7)(128)(4)+(7)(128)(8)+(7)(128)(7)=20608$
- $N=N_X+N_B/100=430$

TST Switch

- TST switch structure

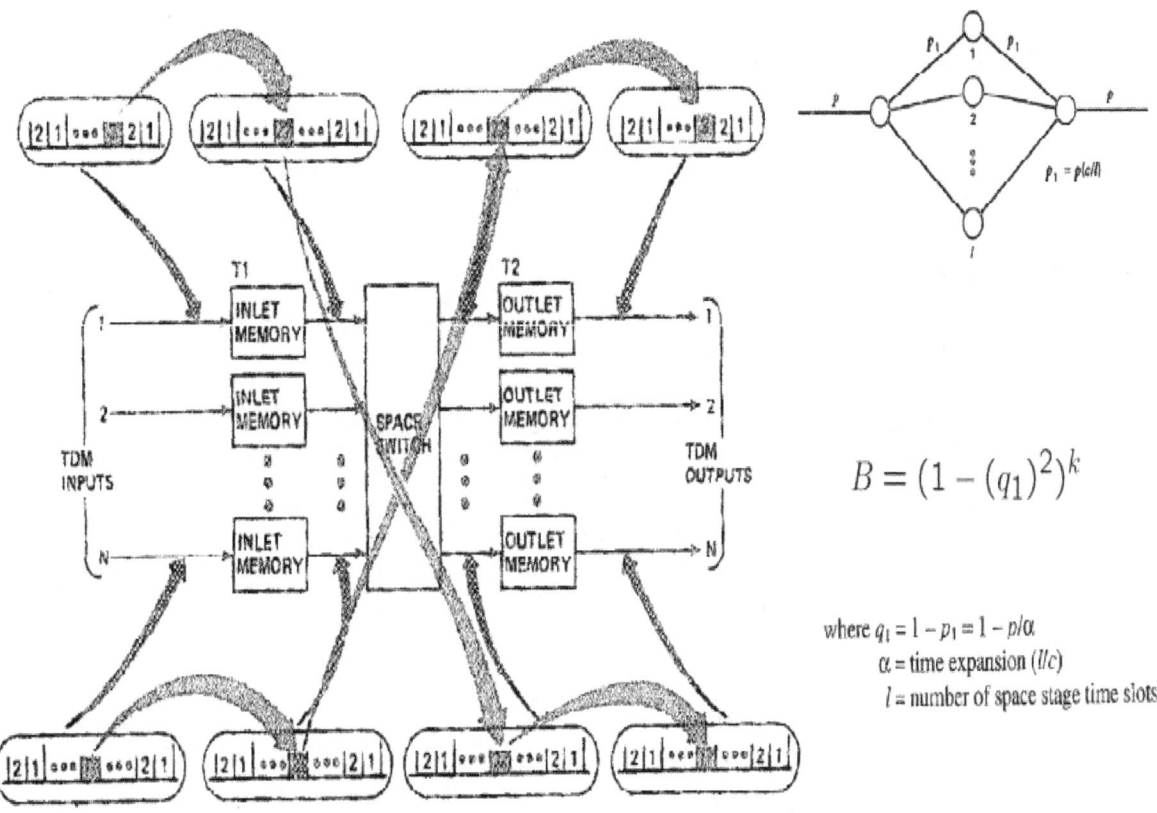

$$B = (1-(q_1)^2)^k$$

where $q_1 = 1 - p_1 = 1 - p/\alpha$
α = time expansion (l/c)
l = number of space stage time slots

TSSST Switching Structure

- Multistage switches

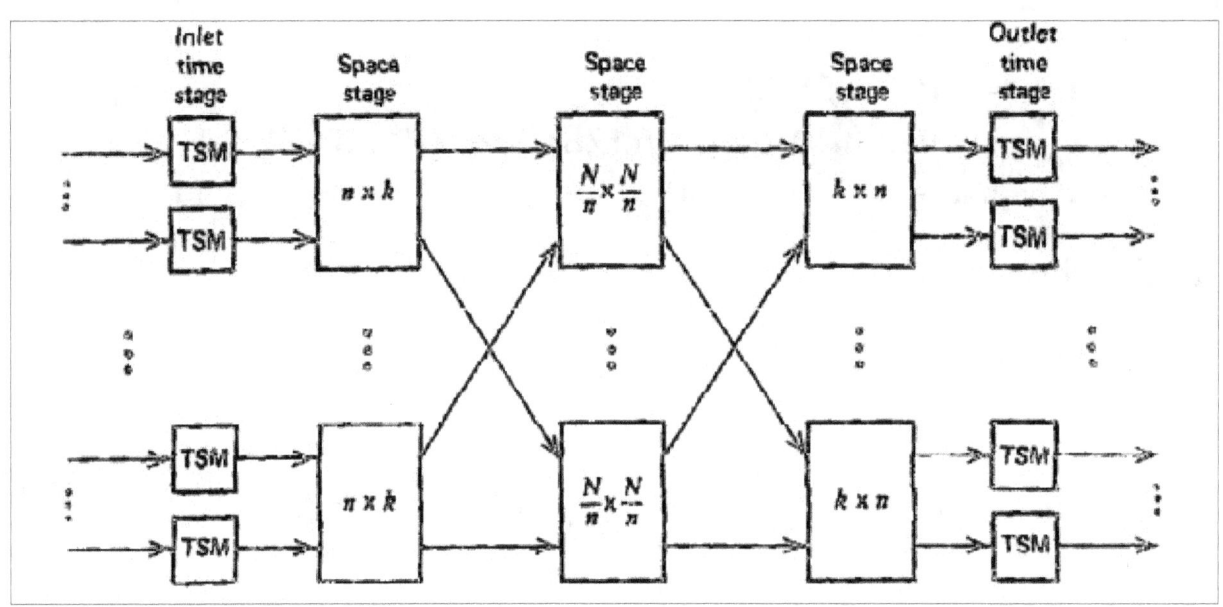

No. 4 ESS Toll Switch

- Electronic Switching System
- Time-space-time with four stages in the space switch

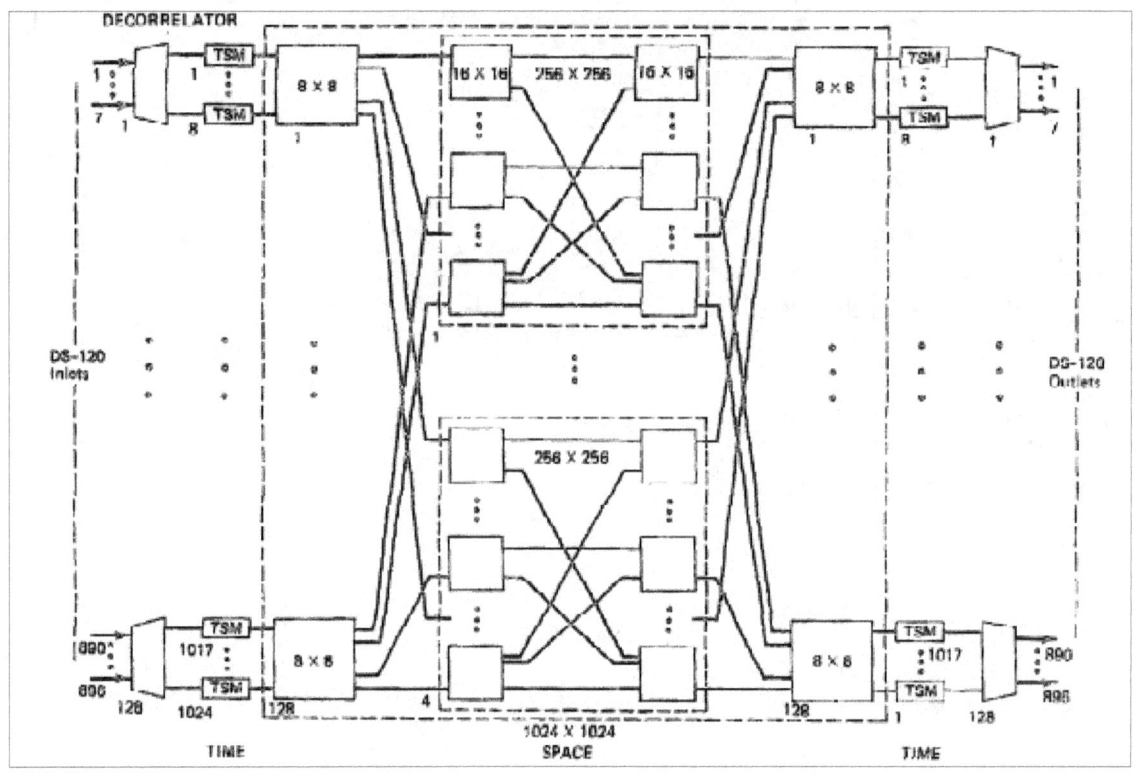

Lecture 4
Backbone Transmission

4.1 Transmission Media

Twisted pair (120 ohm): The physical characteristics of the transmission media used in modern networks. Copper cable pairs has been adopted throughout the World as the means of providing the local loop between the subscriber and the serving telephone exchange. But the humble copper pair cable can also be used to carry a variety of data high speed services, and even video, by the addition of appropriate electronics (e.g. ADSL). Copper pair cable is available as a single pair of wires, with plastic insulation, usually deployed between a subscriber's premises and the overhead or underground distribution point (DP), the copper pairs are also provided in multi-pair cable bundles of a variety of sizes (e.g. from 2 pairs up to 4,800 pairs). The thickness or gauge of the copper conductors also comes in a range of sizes, with corresponding electrical resistance values per kilometre. The individual pairs have a thin insulating cover of polyethylene and these are grouped into quads in order to minimize adverse electrical interference problems. The coating of the pairs is usually color coded to assist their identification by the technician joining the ends onto other cables and terminals – quite a task when there are tens, hundreds or even thousands of pairs in one sheath! The sheath is constructed from polyethylene for strength and durability, together with aluminum foil to provide a water barrier. special forms of copper cable, arranged with one central conductor surrounded by a copper cylindrically shaped conductor – known as 'transverse screen' cable – were deployed in the United Kingdom and elsewhere during the 1970s and 1980s to provide high-speed digital circuits in the access network. Now, optical fiber or digital transmission over standard copper pairs is used in preference.

for the otherwise very fragile bundle of optical fibers. Typical dimensions of the optical fiber's diameters are 50–200 μm core with 125–400 μm cladding for the earlier (multi-mode) types to 5–10 μm core with 125 μm cladding for the more recent (single mode) types. (A 'μm' is one millionth of a meter.) These fine glass threads, which are about the thickness of a human hair, are wrapped in a polythene protective cover for robustness.

- Shielded twisted-pair cable (STP) combines the techniques of shielding, cancellation, and twisting of wires.
- Each pair of wires is wrapped in metallic foil.
- The four pairs of wires are wrapped in an overall metallic braid or foil.
- STP affords greater protection from all types of external interference, but is more expensive and difficult to install than UTP.
- The metallic shielding materials in STP need to be grounded at both ends.

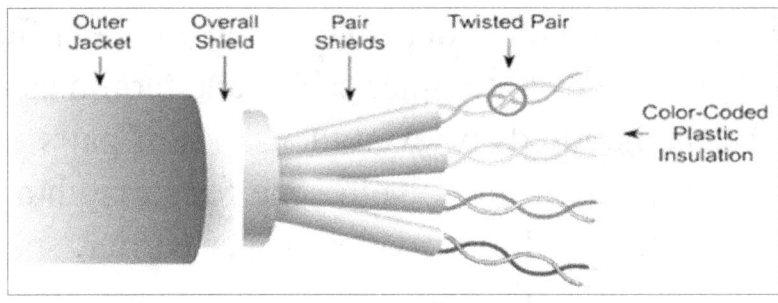

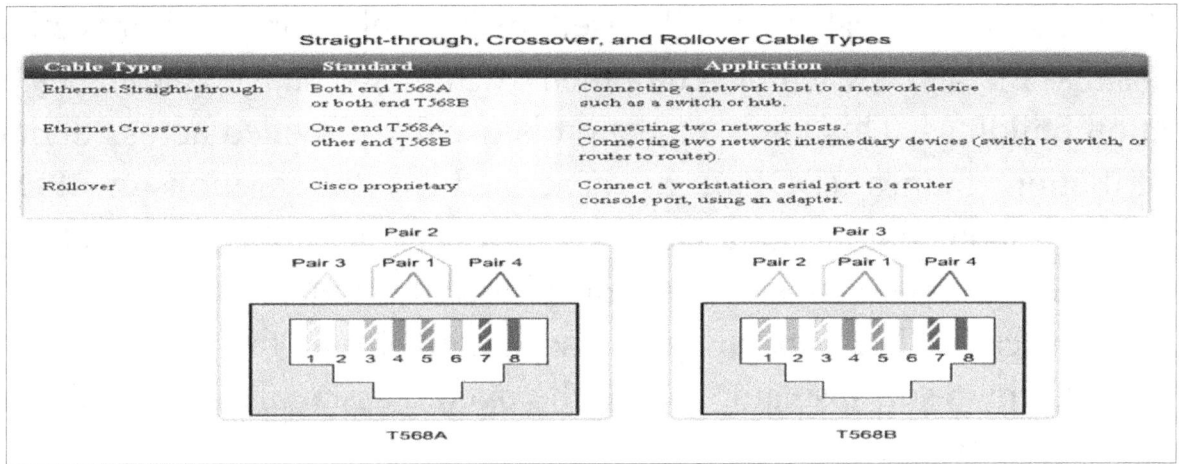

Coaxial cable (75 ohm): As the name suggests, coaxial cable comprises a metallic central core cable encased by a cylindrical conductor. The shielding effect of the outer conductor provides an interference-free transport medium for high-speed electrical signals. Coaxial cable, similar to that used for connecting aerials to a TV set, is used in the Cable TV networks to provide the video connection from the street electronics to an individual household. Early forms of long-distance transmission networks for the PSTN employing analogue FDM systems were provided over coaxial cables from the 1960s to the 1980s.

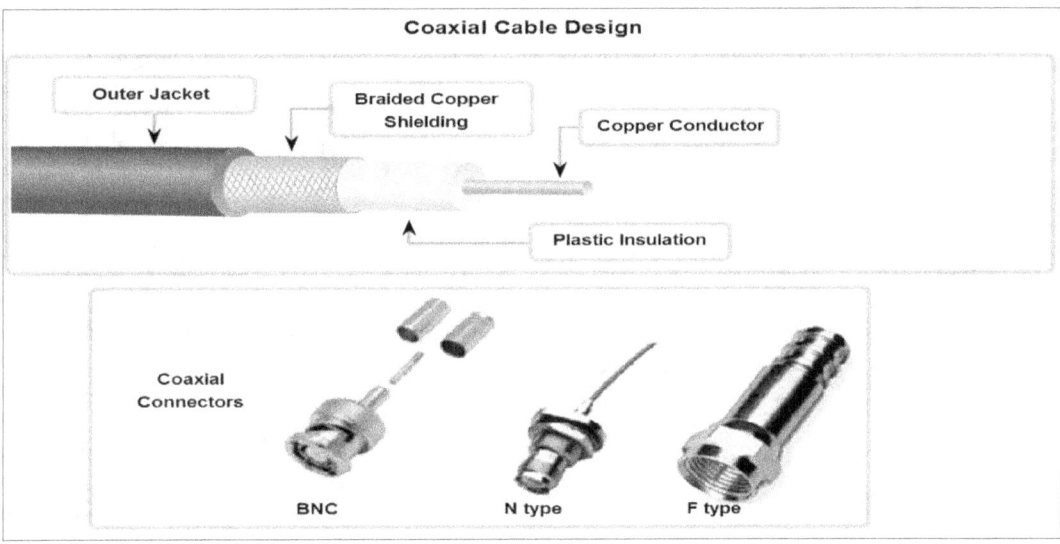

Optical Fiber: Now, telecommunications networks use optical fiber in preference to coaxial cables for the core transmission network. Even though it is not now deployed in the external telecommunications networks, coaxial cable is still used extensively for interconnecting transmission and switching equipment within exchange and repeater station buildings. This interconnection is usually provided across digital distribution frames (DDFs), as described in the section on PDH (pleisochronous digital hierarchy). For physical flexibility the shielding or screening conductor is made out of metallic braiding underneath a plastic sheath. The screening is separated from the central core conductor by a soft insulator so that the overall assembly of the cable is robust and flexible.

- The part of an optical fiber through which light rays travel is called the core of the fiber.

- If the diameter of the core of the fiber is large enough so that there are many paths that light can take through the fiber, the fiber is called "multimode" fiber.

- Single-mode fiber has a much smaller core that only allows light rays to travel along one mode inside the fiber.

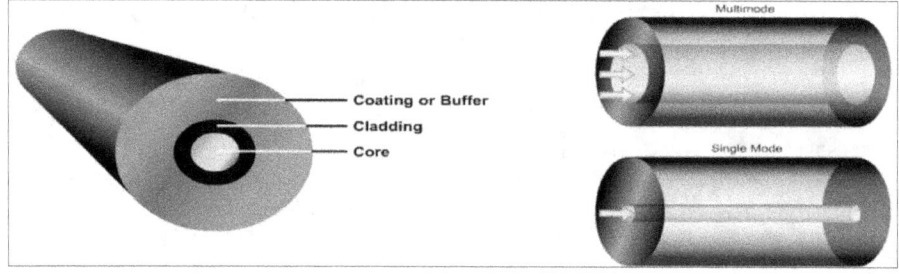

- Fiber-optic cable is not affected by the sources of external noise that cause problems on copper media because external light cannot enter the fiber except at the transmitter end.

- Furthermore, the transmission of light on one fiber in a cable does not generate interference that disturbs transmission on any other fiber.

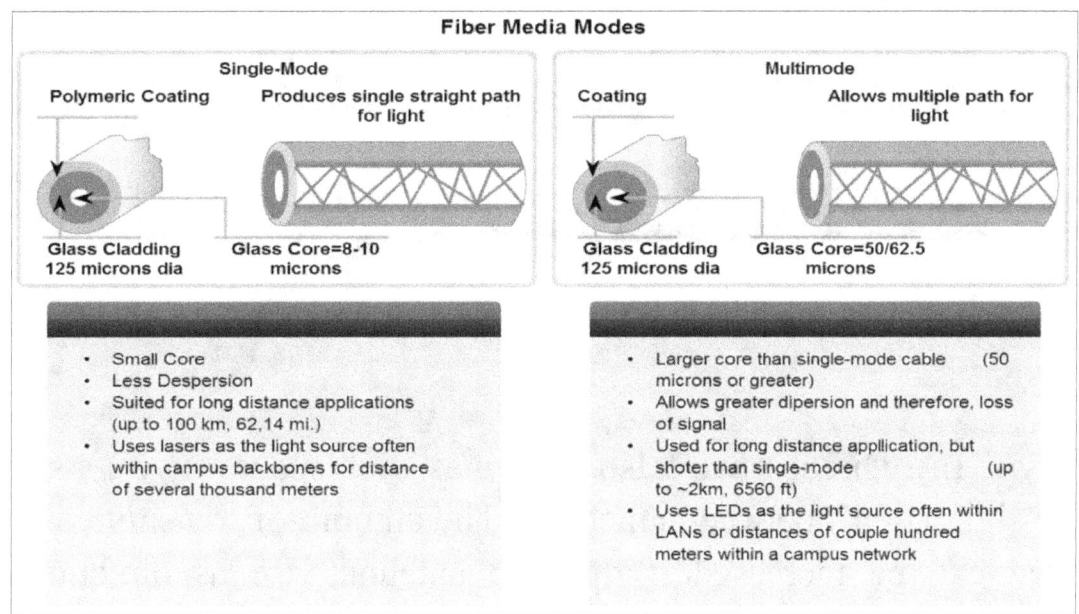

Atmosphere: radio: The medium of the atmosphere is unlike that of the confined environment of metallic cables or optical fibers, and its unbounded nature requires the transmission systems exploiting it to have a high degree of adaptability to the medium's changing characteristics. Radio transmission through the air is achieved in several ways, mainly

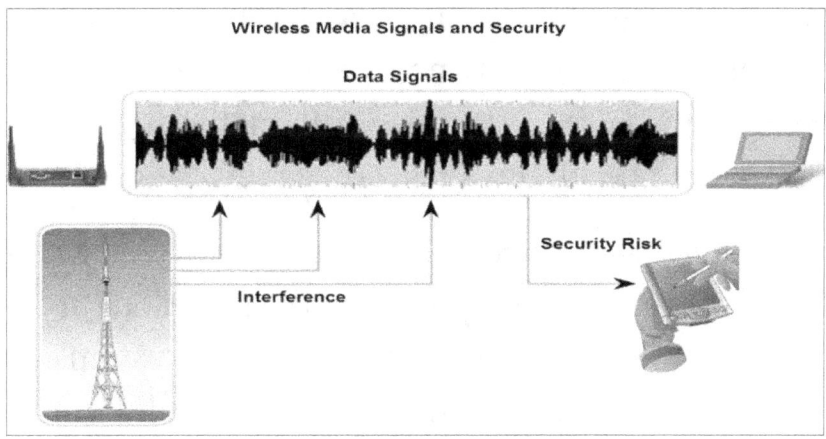

depending on the frequency of the carrier wave used.

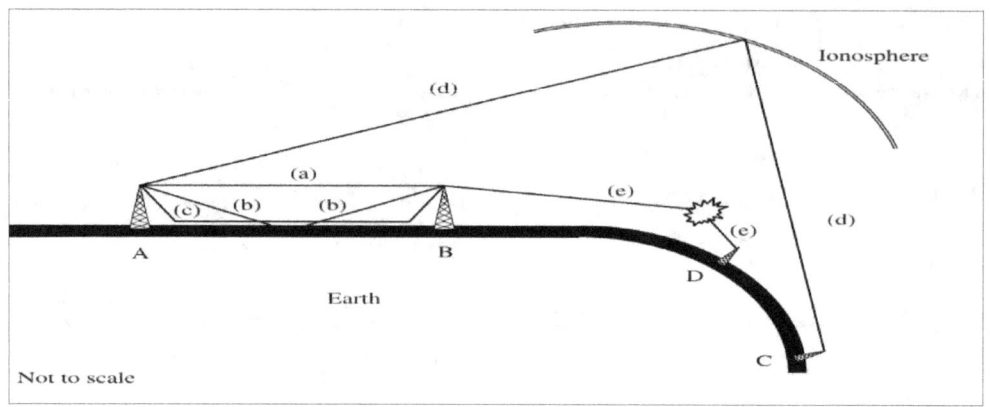

Path type (a): Direct wave (also known as 'free space wave') a straight line-of-sight (LOS) between antennas. The amount of transmitted power received can be increased if the two antennas are as directional as possible. Since the directional characteristics of antennas increase with frequency, practical systems tend to be at microwave frequencies, that is above 1 GHz (1,000,000,000 Hz). Path type (b): Ground reflected wave in addition to the direct wave there will inevitably be one or more reflected wave paths between the sending and receiving antennas. Normally, these reflected paths cause some degree of interference with the direct signal at the receiver because the reflected signal will experience greater delay, and hence be out of phase with the direct signal. LOS systems use a variety of methods for reducing the interference from the reflected waves.

Path type (c) ground or surface wave. Ground waves, which follow the surface of the Earth through the process of diffraction, are generated by electrical currents induced in the ground. Radio systems using surface waves are able to extend beyond the LOS, following the Earth's curvature. The attenuation of surface waves increases with frequency, so practical systems operate in the low frequency range (30–30 kHz) and, thus, are of low capacity.

Path type (d): skywave: This path is created by the reflection of the radio waves off the Ionosphere, the ionized layers belting the Earth. The waves can carry over very long distances through multiple reflections off the Ionosphere and ground. Early long distant, i.e. intercontinental, communications were achieved using skywave radio systems operating at HF frequencies. However, the variable nature of the Ionosphere and the use of multi-path propagation meant that the transmission was subject to fading and daily fluctuations. Thus, alternative transmission systems (e.g. optical fiber, microwave radio and Earth satellite) are mainly used in preference to HF.

Path type (e): tropospheric scattering. This path, which is created by the scattering of the radio wave by perturbations in the atmosphere, enables communication to a receiver just over the horizon. Practical systems operate at microwave frequencies using very high transmitting power a large high gain receiving antennas to compensate for the very high attenuation of tropospheric scattering paths. Such systems are used by the military to provide communication in battle conditions. BT use tropospheric systems to connect all the oil rigs in the North Sea to mainland UK, thus extending the PSTN offshore beyond the horizon.

the configuration for a microwave point-to-point relay route, providing transmission capacity between two antennas via an intermediate relay stage (repeater or amplifier for digital or analogue transmission, respectively). The Go and Return channels are conveyed on separate carrier frequencies. Such transmission systems are able to provide routes across a country using many repeater antennas, typically spaced at between 20–40 km apart.

Free space: Earth satellites: By locating the microwave radio relay in a communications satellite in space orbit around the World, the distance between a pair of antennas can be extended to intercontinental distances. Clearly, the free space loss of the radio signal is correspondingly large because of the long distances between Earth antennae and the satellite relay – typically this so-called path loss up and down totals around 200 dB, depending of the frequency of the carrier wave and the path length, the latter depending on the latitude of the Earth stations.

Satisfactory performance requires compensation for this loss by the use of high-gain repeaters and antennas on board the satellite, as well as high-gain antennas on the ground. Since the gain of an antenna increases with size, the gain of the antennas on board the satellite is limited by the weight and diameter of the antennas that can be mounted on the satellite, so the gain of the antennas on the ground need to be very high – hence the use of the huge satellite dish aerials, which are a familiar sight at satellite Earth stations.

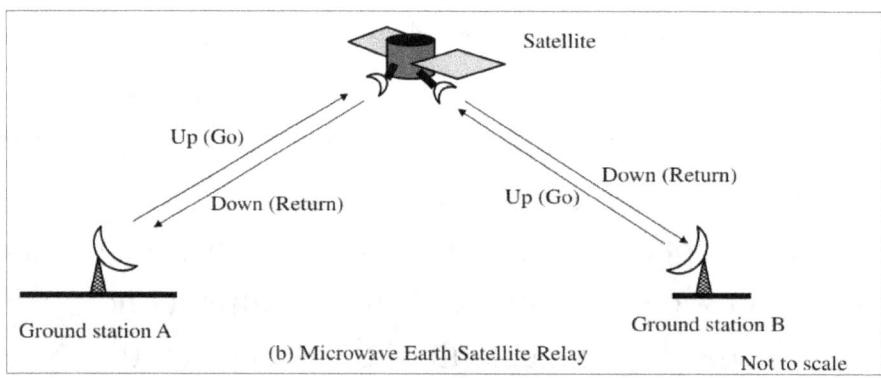

(b) Microwave Earth Satellite Relay

Alternatively, smaller capacity transmission links can be achieved from a central large dish antenna at one end of the system and the use of small portable antennas at the other, e.g. small aperture satellite systems as used by news TV broadcasters, and hand-held satellite mobile phones used in remote areas of the World

4.2 Digital Transmission systems

Unlike the Access Network, which may link some tens of thousands of subscribers to a single serving local exchange, the Core Transmission Network provides links between relatively small numbers of network nodes, typically spread across the whole country. These transmission nodes are points in the national network where bundles of circuits, serving telephony, private circuits, data services, etc., are extracted from or entered onto the required transmission links. Economies of scale are achieved by multiplexing onto as few, but large, transmission systems as possible – since the unit costs of transmission systems decreases with system size. Thus, in many cases it is economical to unpack and re-pack transmission links at intermediate points within a network in order to achieve optimum loading of transmission links.

4.2.1 Pleisochronous Digital hierarchy (PDH)

The practical arrangement at a CTS is shown in the next figure, in which a 2 Mbit/s block is extracted from the incoming 140 Mbit/s PDH transmission system on the left and inserted into the outgoing 140 Mbit/s PDH transmission system on the right (and vice versa for the return direction of transmission). The CTS is composed of racks of transmission terminal equipment and three stages of multiplexors:

140/34, 34/8 and 8/2 – each connected to a common DDF, or to one of several DDFs. In this example, the incoming 140 Mbit/s system terminates on 140/34 Mbit/s mux. No. 1, from which a jumper wire is run from output No.1 to the input of 34/8 Mbit/s mux. No. 1, and so on via the central DDF to the output multiplexor mountain through

to 140/34 Mbit/s mux. No. 11. The figure also shows the extraction at the DDF of 2 Mbit/s blocks from the input 140 Mbit/s transmission system (via 34/8 Mbit/s muxs No. 1 and No. 4, and 8/2 Mbit/s muxs No. 1 and No. 13) to the co-sited telephone switching unit. Transmission flexibility is also possible at any of the intermediate transmission rates of 34 and 8 Mbit/s through appropriate jumpering at the DDFs. Such a facility would be used, for example, to route a 34 Mbit/s digital private circuit through

the CTS. The jumper wires on the DDFs are coaxial type cables, similar to those used to connect the domestic TV to its aerial. Each of the jumperings have to be made manually when the routes are set up. Not only does this represent a current account cost for the network operator, but the periodic intervention on the DDFs to add or remove jumper wires also introduces disturbance to the established jumpers, often resulting in damage and faults. Electronic replacements for the jumper wires and DDFs, using digital cross-connect equipment, known as 'DXC' in the USA, are used by some network operators. However, for cost reasons most network operators tend to use DXC to provide the transmission flexibility only for the high value international transmission systems at the larger international gateway CTSs.

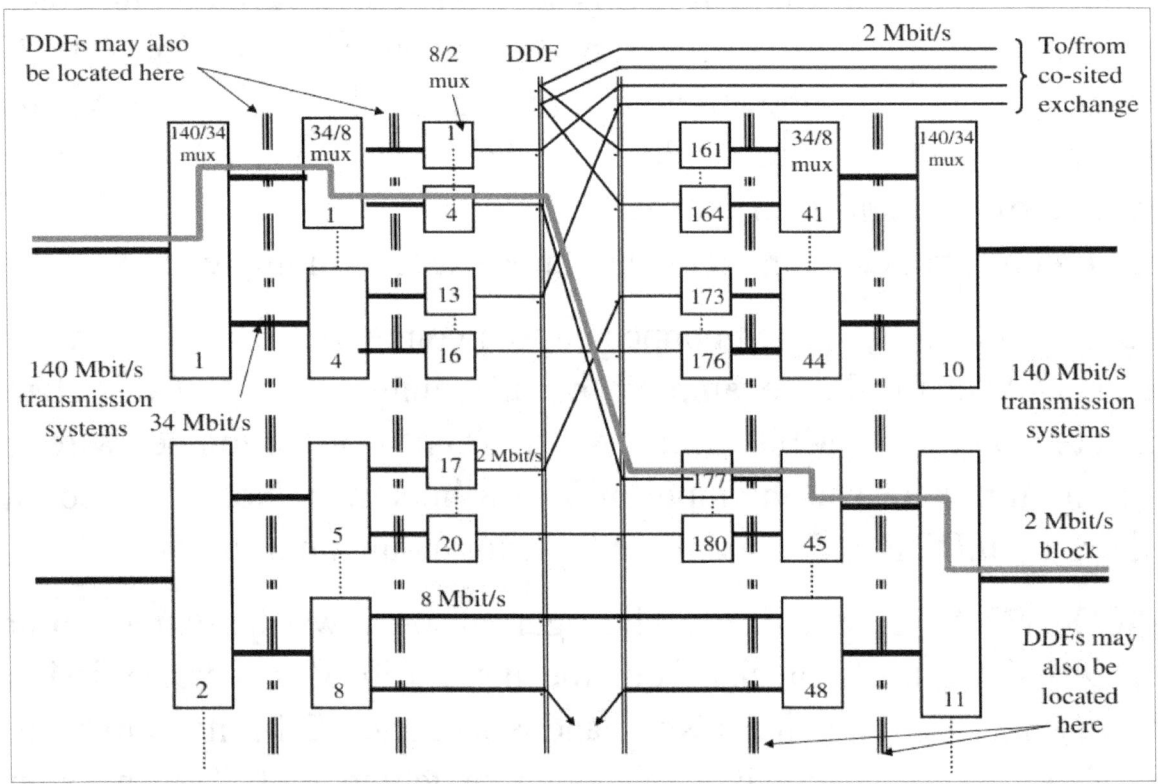

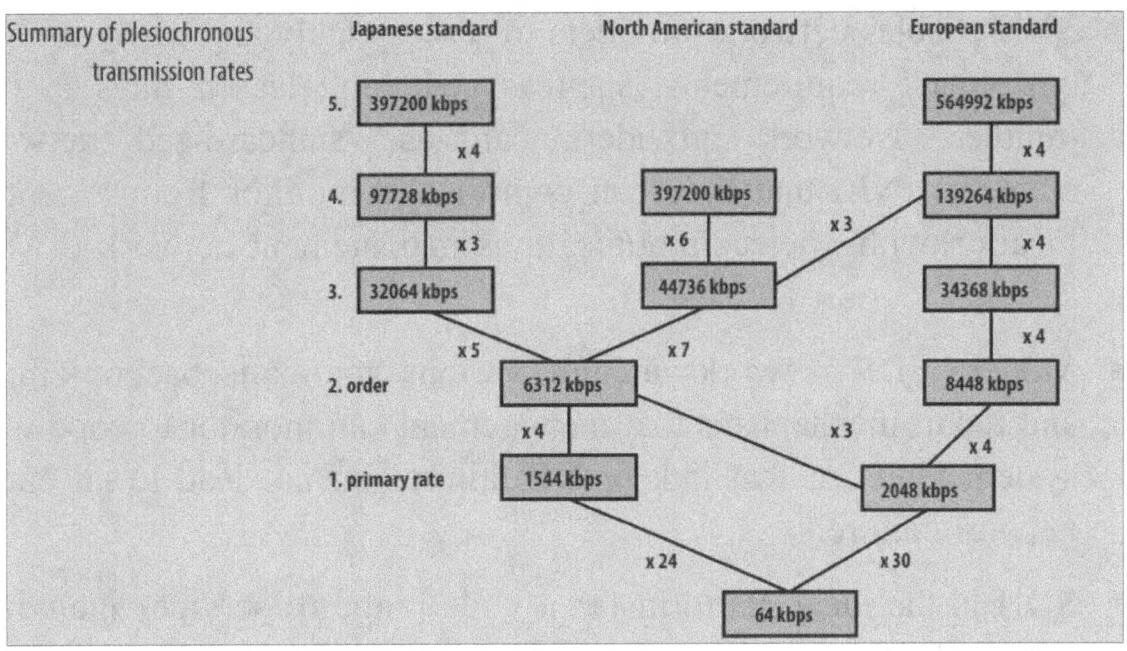

4.2.2 Synchronous Digital Hierarchy (SDH)

Communications networks gradually converted to digital technology after PCM was introduced in the 1960s. A multiplex hierarchy known as pleisochronous digital hierarchy (PDH) evolved to cope with the demand for ever-higher bit rates. The bit rates start with the basic multiplex rate of 2 Mbps with further stages of 8, 34, and 140 Mbps. In North America and Japan, however, the primary rate is 1.5 Mbps with additional stages of 6 and 44 Mbps, as shown in Figure 1 on page 5. This fundamental developmental difference made gateway setup between the networks both difficult and expensive. In response to the demand for increased bandwidth, reliability, and high-quality service, SDH developed steadily during the 1980s, eliminating many inherent disadvantages in PDH. Transmission rates of up to 10 G can be achieved in modern SDH systems making it the most suitable technology for backbones, the superhighways in today's telecommunications networks.

- Compared to the older PDH system, low-bit-rate channels can be easily extracted from and inserted into the high-speed bit streams in SDH, eliminating the need for Costly demultiplexing and re-multiplexing the pleisochronous structure.

- SDH enables network providers to react quickly and easily to their customers' requirements, such as switching leased lines in just minutes. Network providers can use standardized network elements (NE) that they can control and monitor from a central location with a telecommunications management network (TMN) system.

- Modern SDH networks include various automatic backup-circuit and repair mechanisms that management can monitor to cope with system faults so that link or NE failures do not lead to an entire network failure.

- SDH is the ideal platform for a wide range of services including POTS, ISDN, mobile radio, and data communications, such as LAN and WAN. It can also handle more recent services such as video on demand and digital video broadcasting via ATM

- SDH simplifies gateway setup between different network providers and to SONET systems. The SDH interfaces are globally standardized, making it possible to combine NEs from different manufacturers into a single network which reduces equipment costs.

- The trend in transport networks is toward ever-higher bit rates such as STM-256 (time division multiplex, TDM); however, these NEs can be cost-prohibitive. Another cheaper alternative is dense wavelength division multiplexing (DWDM) which uses multiple single-mode optical fibers to transmit digital signals through several fibers simultaneously.

- DWDM is closely associated with the "all-optical network." In terms of the ISO-OSI layer model, OADM and optical cross-connect (OXC) development basically means introducing an additional DWDM layer below the SDH layer. Therefore, future systems will likely combine higher multiplex rates with DWDM

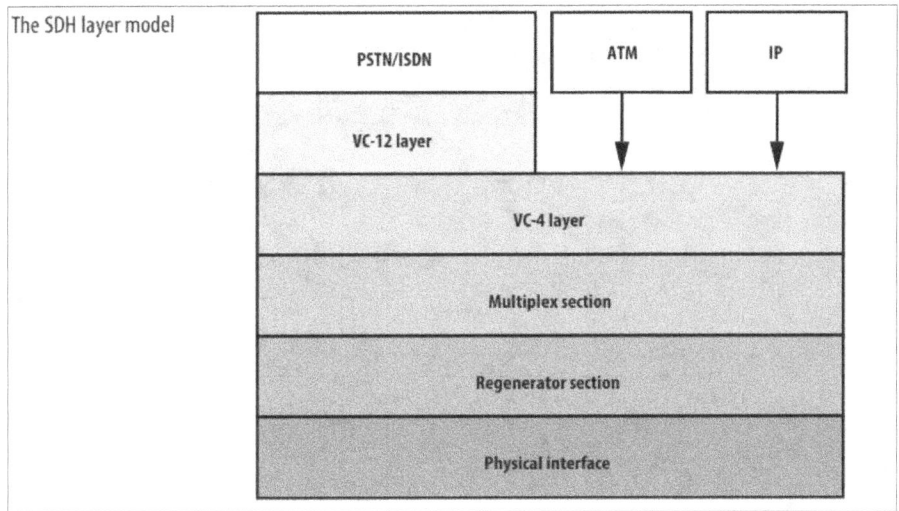

SDH networks are subdivided into various layers directly related to the network topology. The lowest layer is the physical layer, which represents the transmission medium, usually a glass fiber or possibly a radio or satellite link. The regenerator section is the path between regenerators, see next Figure. Part of the regenerator section overhead (RSOH) is available for the signaling required within this layer.

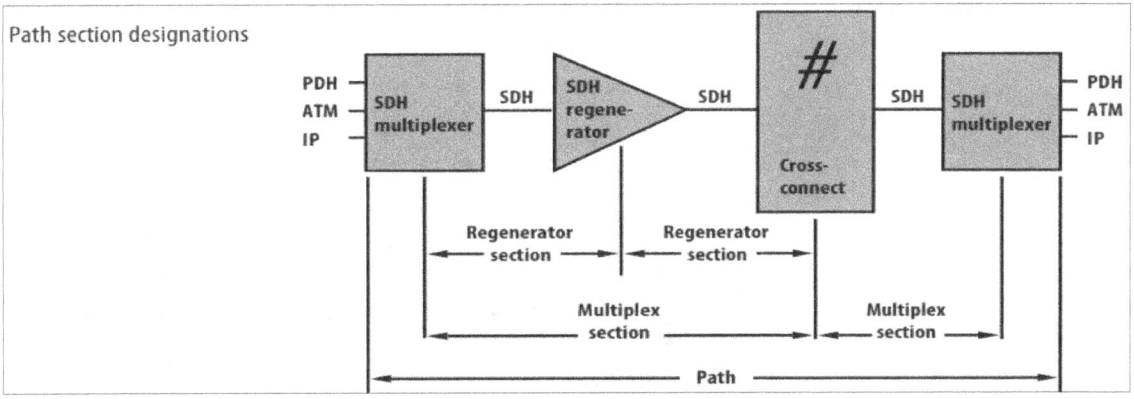

The remainder of the overhead, the multiplex section overhead (MSOH) is used for multiplex section needs and covers the part of the SDH link between multiplexers. The carriers or virtual containers (VC) are available as payload at each end of this section which represent a part of the mapping process, the procedure of packing the tributary signals such as PDH and ATM into SDH transport modules. VC-4 mapping is used for 140 Mbps or ATM signals and VC-12 mapping is used for 2 Mbps signals.

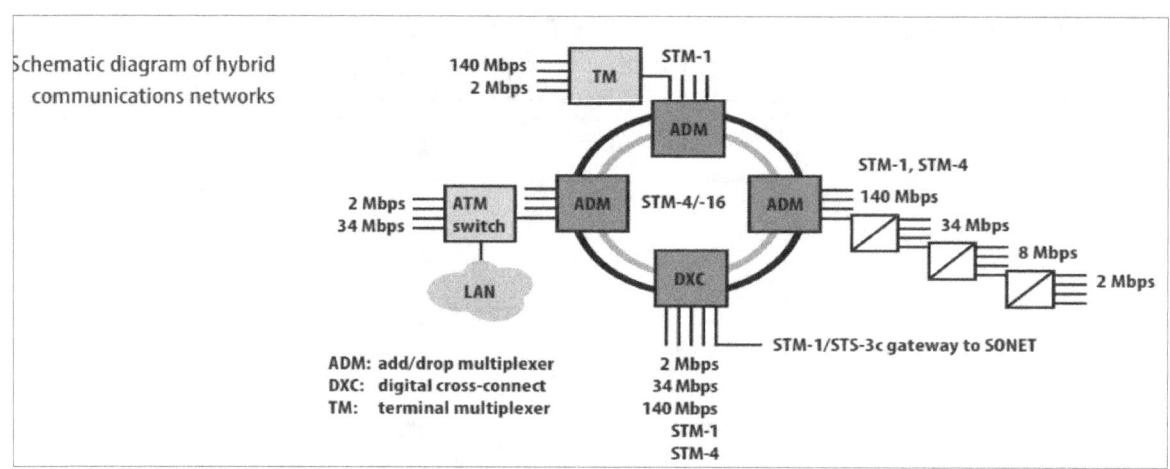

Schematic diagram of hybrid communications networks

ADM: add/drop multiplexer
DXC: digital cross-connect
TM: terminal multiplexer

A 155.52 Mbps frame defined in ITU-T recommendation G.707 is known as the synchronous transport module (STM). Since this frame is the first level of the synchronous digital hierarchy, it is known as STM-1. It comprises a byte matrix of 9 rows and 270 columns. Transmission is row by row, starting with the byte in the upper-left corner and ending with the byte in the lower-right corner. The frame repetition rate is 125 ms. Each byte in the payload represents a 64 kbps channel. The STM-1 frame can transport any PDH tributary signal ($\leq$ 140 Mbps).

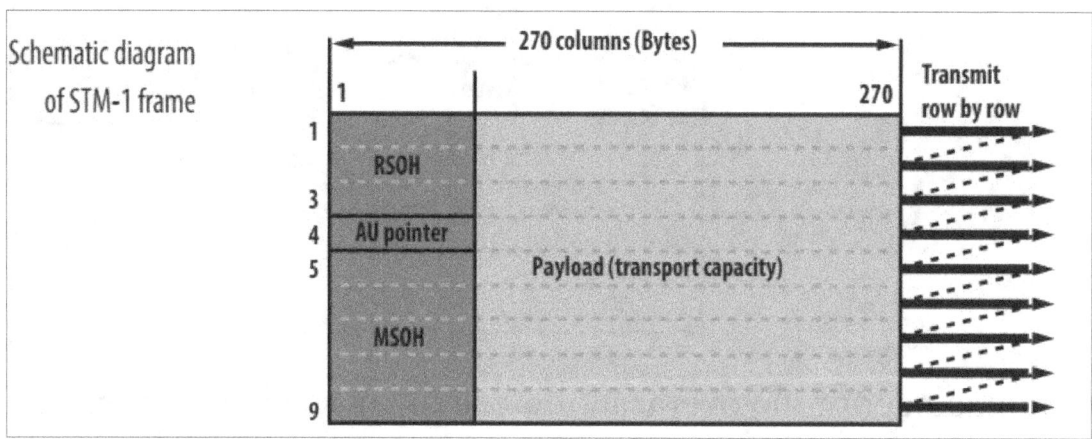

Schematic diagram of STM-1 frame

The first nine bytes in each of the nine rows are called the section overhead (SOH), G.707 makes a distinction between the regenerator

section overhead (RSOH) and the multiplexer section overhead (MSOH) so that the functions of certain overhead bytes can be coupled with the network architecture.

Overview of STM-1 overhead

A1	A1	A1	A2	A2	A2	J0	X	X
B1	-	-	E1	-	-	F1	X	X
D1	-	-	D2	-	-	D3		
AU pointer								
B2	B2	B2	K1			K2		
D4			D5			D6		
D7			D8			D9		
D10			D11			D12		
S1						M1	E2	

X   Reserved for national use
-    Media-dependent use (radio-link, satellite)

Overhead bytes and their functions

Overhead Byte	Function
A1, A2	Frame alignment
B1, B2	Quality monitoring, parity bytes
D1 ... D3	QECC network management
D4 ... D12	QECC network management
E1, E2	Voice connection
F1	Maintenance
J0 (C1)	Trace identifier
K1, K2	Automatic protection switching (APS) control
S1	Clock quality indicator
M1	Transmission error acknowledgment

VC-3/4 POH

Byte	Function
J1	Path indication
B3	Quality monitoring
C2	Container format
G1	Transmission error acknowledgment
F2	Maintenance
H4	Superframe indication
F3	Maintenance
K3	Automatic protection switching
N1	Tandem connection monitoring

The VC-3/4 POH is the high-order path overhead for transporting 140 Mbps, 34 Mbps, and ATM signals.

VC-11/12 POH

Byte	Function
V5	Indication and error monitoring
J2	Path indication
N2	Tandem connection monitoring
K4	Automatic protection switching

The VC-11/12 POH is the low-order path for transporting 1.544 Mbps, 2.048 Mbps, and ATM signals.

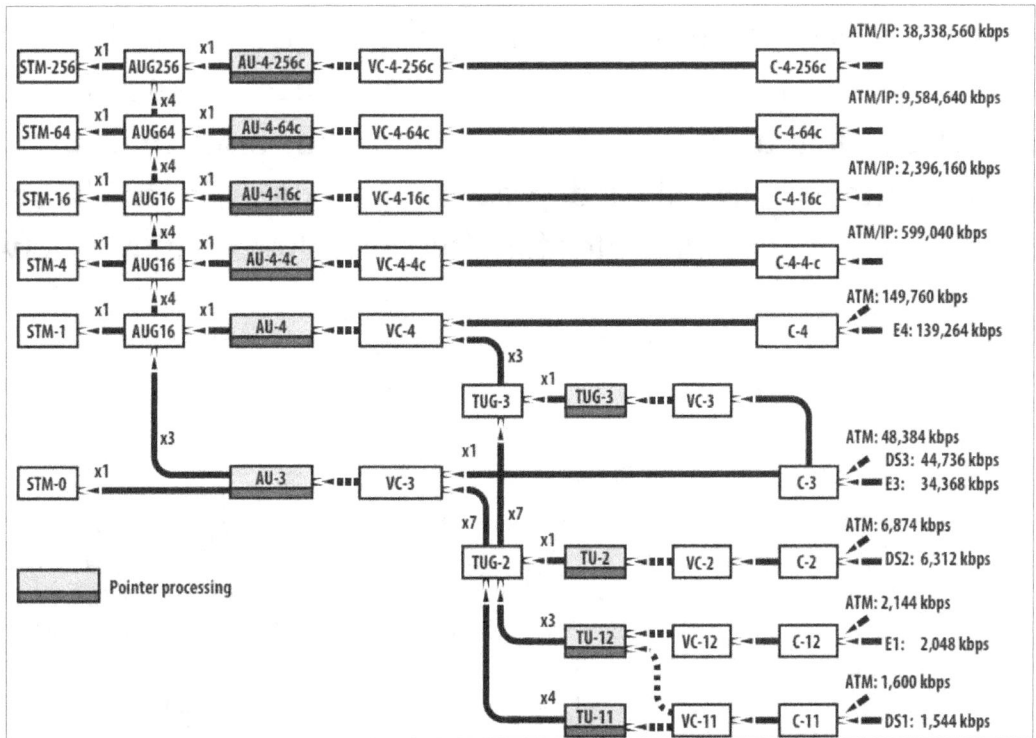

SONET/SDH signal and bit-rate hierarchy	SONET Signal		Bit Rates (Mbps)	Equivalent SDH Signal
	STS-1	OC-1	51.84	STM-0
	STS-3	OC-3	155.52	STM-1
	STS-12	OC-12	622.08	STM-4
	STS-48*	OC-48	2488.32	STM-16
	STS-192*	OC-192	9953.28	STM-64
	STS-768	OC-768	39813.12	STM-256

The use of pointer procedures also gives synchronous communications a distinct advantage over the pleisochronous hierarchy. Pointers are used to localize individual virtual containers in the payload of the synchronous transport module. The pointer may directly indicate a single VC-n virtual container from the upper level of the STM-1 frame. Chained pointer structures can also be used. The AU-4 pointer initially indicates the VC-4 overhead. Three further pointers are located at fixed positions in the VC-4 and are used to indicate the start of the three VC-3 virtual containers relative to the VC-4.

Next Figure illustrates the pointer procedure using C3 mapping as an example. SDH multiplexers are controlled from a highly accurate central clock source running at 2.048 MHz. Pointer adjustment may be necessary if phase variations occur in the actual network; or when feeding the connection through various provider networks, the AU pointer can be altered in every fourth frame with prior indication. The virtual container is then shifted precisely by three bytes. Therefore, pointer activity is a good indicator for clock variations within a network.

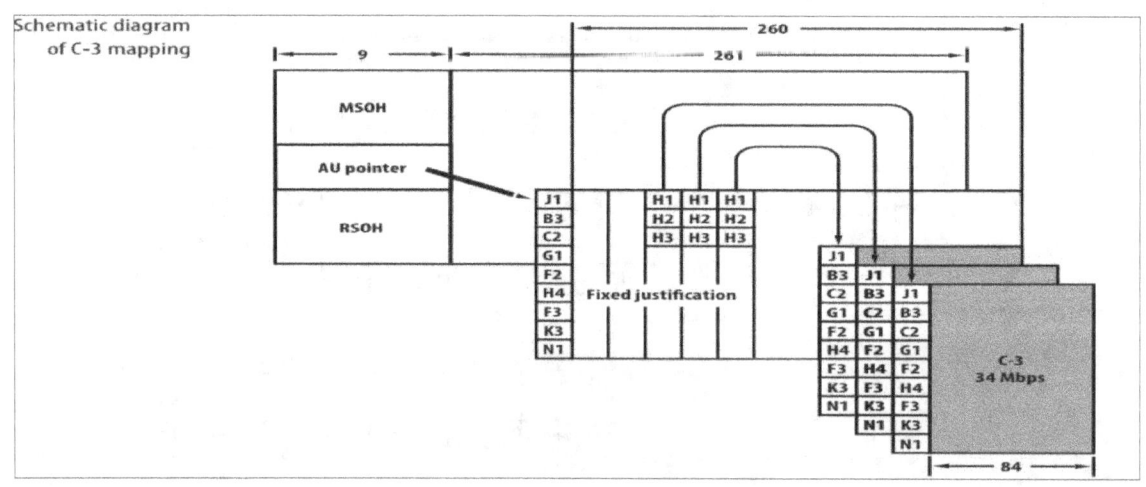

Negative pointer justification	
	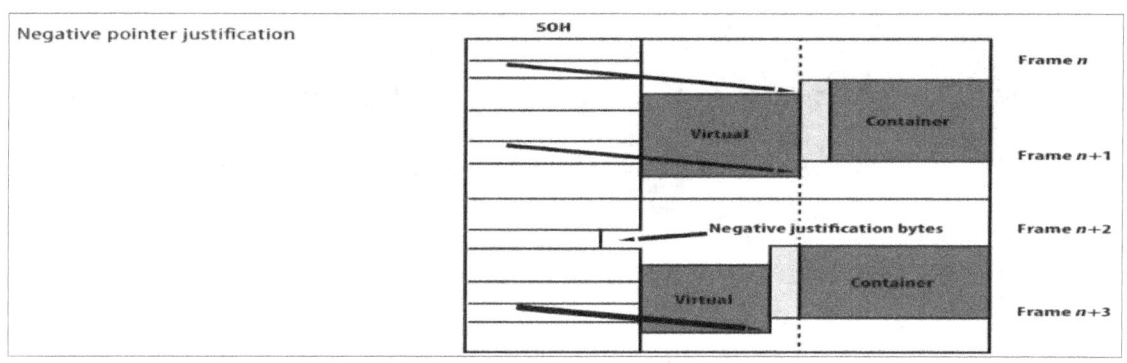

Overview of major defects and anomalies	
	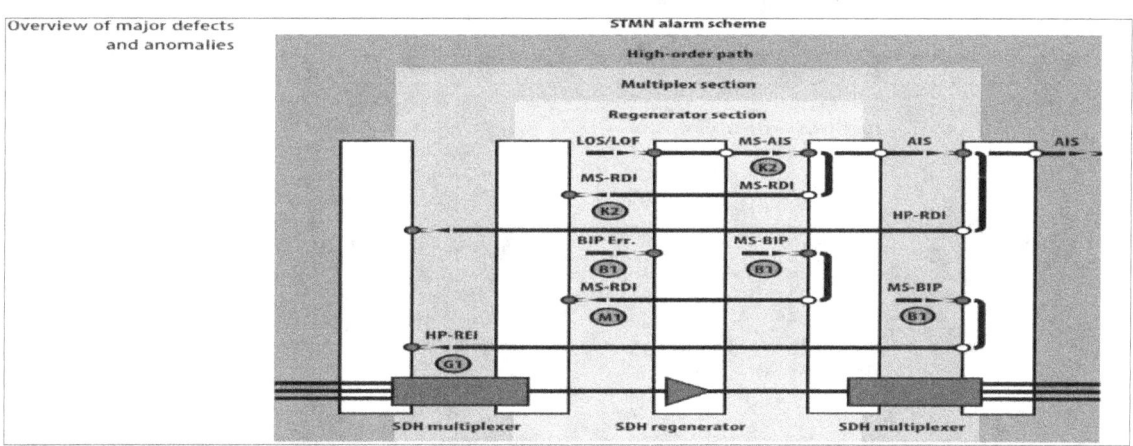

Anomalies and defects in SDH		**Anomalies/Defects**	**Detection Criteria**
	LOS	Loss of signal	A drop in incoming optical power level causes a high bit error rate
	OOF	Out of frame	A1, A2 errored for ≥ 625 μs
	LOF	Loss of frame	If OOF persists for ≥ 3 ms
	RS BIP Error	Regenerator section BIP error (B1)	Mismatch of the recovered and computed BIP-8 Covers the whole STM-N frame
	RS-TIM	Regenerator section trace identifier mismatch	Mismatch of the accepted and expected trace identifier in byte J0
	MS BIP error	Multiplex section BIP error (B2)	Mismatch of the recovered and computed $N \times$ BIP-24 Covers the whole frame except RSOH
	MS-AIS	Multiplex section alarm indication signal	K2 (bits 6, 7, 8) = 111 for 3 frames

	Anomalies/Defects	**Detection Criteria**
MS-REI	Multiplex section remote error indication	Number of detected B2 errors in the sink side encoded in byte M1 of the source side
MS-RDI	Multiplex section remote defect indication	K2 (bits 6, 7, 8) = 111 for $\geq z$ frames ($z = 3$ to 5)
AU-AIS	Administrative unit alarm indication signal	All ones in the AU pointer bytes H1 and H2
AU-LOP	Administrative unit loss of pointer	8 to 10 NDF enable 8 to 10 invalid pointers
HP BIP error	HO path BIP error (B3)	Mismatch of the recovered and computed BIP-8 Covers entire VC-n
HP-UNEQ	HO path unequipped	C2 = 0 for ≥ 5 frames
HP-TIM	HO path trace identifier mismatch	Mismatch of the accepted and expected trace identifier in byte J1
HP-REI	HO path remote error indication	Number of detected B3 errors in the sink side encoded in byte G1 (bits 1, 2, 3, 4) of the source side
HP-RDI	HO path remote defect indication	G1 (bit 5) = 1 for $\geq z$ frames ($z = 3, 5,$ or 10)

	Anomalies/Defects	Detection Criteria
HP-PLM	HO path payload label mismatch	Mismatch of the accepted and expected Payload Label in byte C2
TU-LOM	Loss of multiframe $x = 1$ to 5 ms	H4 (bits 7, 8) multiframe not recovered for X ms
TU-AIS	Tributary unit alarm indication signal	All ones in the TU pointer bytes V1 and V2
TU-LOS	Tributary unit loss of pointer	8 to 10 NDF enable 8 to 10 invalid pointers
LP BIP error	LO path BIP error	Mismatch of the recovered and computed BIP-8 (B3) or BIP-2 (V5 bits 1, 2) Covers entire VC-n
LP-UNEQ	LO path unequipped	VC-3: C2 = 0 for $\geq$ 5 frames VC-m ($m = 2, 11, 12$): V5 (bits 5, 6, 7) = 000 for $\geq$ 5 multiframes
LP-TIM	LO path trace identifier mismatch	Mismatch of the accepted and expected trace identifier in byte J1 (VC-3) or J2

	Anomalies/Defects	Detection Criteria
LP-REI	LO path remote error indication	VC-3: Number of detected B3 errors in the sink side encoded in byte G1 (bits 1, 2, 3, 4) of the source side VC-m ($m = 2, 11, 12$): If one or more BIP-2 errors detected in the sink side, byte V5 (bits 3) = 1 on the source side
LP-RDI	LO path remote defect indication	VC-3: G1 (bit 5) = 1 for $\geq z$ frames VC-m ($m = 2, 11, 12$): V5 (bit 8) = 1 for $\geq z$ multiframes ($z = 3, 5,$ or 10)
LP-PLM	LO path payload label mismatch	Mismatch of the accepted and expected payload label in byte C2 or V5 (bits 5, 6, 7)

Synchronization: Failure to guarantee synchronization can result in considerable degradation in network functionality and sometimes even total failure. To avoid this, synchronize all NEs to a central clock generated by a highly precise, primary reference clock (PRC) that conforms to ITU-T recommendation G.811 which specifies 1x10-11accuracy. The clock signal must then be distributed throughout the network. This hierarchical structure passes the signal on through subordinate synchronization supply units (SSU) and synchronous equipment clocks (SEC) using the same synchronization signal paths as those used for SDH communications.

Clock supply hierarchy structure

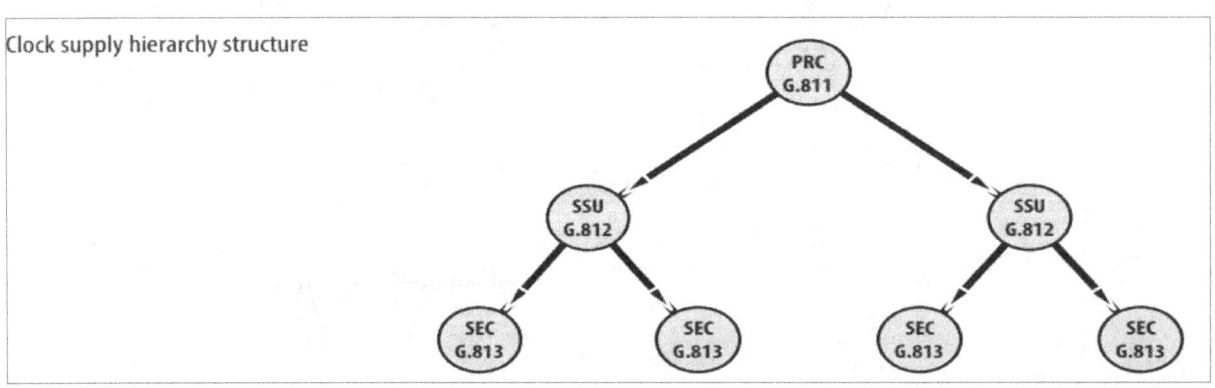

The clock signal is regenerated in the SSUs and SECs with the aid of phase-locked loops. If the clock supply fails, the affected NE switches over to a clock source with the same or lower quality. If a clock source is unavailable, the NE switches to holdover mode which keeps the clock signal relatively accurate by controlling the oscillator, applying stored frequency correction values for the previous hours, as well as considering the oscillator's temperature. Avoid clock "islands" at all costs, as they will eventually drift out of synchronization leading to a total failure. These islands can be prevented using synchronization status messages (SSM – part of the S1 byte) to signal the NEs. The SSM informs the neighboring NE of the clock supply status as part of the overhead. Certain problems can arise at the gateways between networks with independent clock supplies; however, SDH NEs can somewhat compensate for clock offsets through pointer operations. Pointer activity can be a reliable indicator of clock supply problems.

4.2.3 Optical Transport Network (OTN)

The OTN architecture concept was developed by the ITU-T initially a decade ago, to build upon the Synchronous Digital Hierarchy (SDH) and Dense Wavelength-Division Multiplexing (DWDM) experience and provide bit rate efficiency, resiliency and management at high capacity. OTN therefore looks a lot like Synchronous Optical Networking (SONET) / SDH in structure, with less overhead and more management features. It is a common misconception that OTN is just SDH with a few insignificant changes. Although the multiplexing structure and terminology look the same, the changes in OTN have a great impact on its use in, for example, a multi-vendor, multi-domain environment. OTN was created to be a carrier technology, which is why emphasis was put on enhancing transparency, reach, scalability and monitoring of signals carried over large distances and through several administrative and vendor domains. All these are issues that the NREN community is currently struggling to solve.

The advantages of OTN compared to SDH are mainly related to the introduction of the following changes:

- **Transparent Client Signals:** In OTN the Optical Channel Payload Unit-k (OPUk) container is defined to include the entire SONET/SDH and Ethernet signal, including associated overhead bytes, which is why no modification of the overhead is required when transporting through OTN. This allows the end user to view exactly what was transmitted at the far end and decreases the complexity of troubleshooting as transport and client protocols aren't the same technology. OTN uses asynchronous mapping and demapping of client signals, which is another reason why OTN is timing transparent.

- **Better Forward Error Correction**: OTN has increased the number of bytes reserved for Forward Error Correction (FEC), allowing a theoretical improvement of the Signal-to-Noise Ratio (SNR) by 6.2 dB. This improvement can be used to enhance the optical systems in the following areas:

  o Increase the reach of optical systems by increasing span length or increasing the number of spans.

  Increase the number of channels in the optical systems, as the required power theoretical has been lowered 6.2 dB, thus also reducing the non-linear effects, which are dependent on the total power in the system.

  o The increased power budget can ease the introduction of transparent optical network elements, which can't be introduced without a penalty. These elements include Optical Add-Drop Multiplexers (OADMs), Photonic Cross Connects (PXCs), splitters, etc., which are fundamental for the evolution from point-to-point optical networks to meshed ones.

- The FEC part of OTN has been utilised on the line side of DWDM transponders for at least the last 5 years, allowing a significant increase in reach/capacity

- Better scalability: The old transport technologies like SONET/SDH were created to carry voice circuits, which is why the granularity was very dense – down to 1.5 Mb/s. OTN is designed to carry a payload of greater bulk, which is why the granularity is coarser and the multiplexing structure less complicated.

- Tandem Connection Monitoring: The introduction of additional Tandem Connection Monitoring (TCM) combined with the decoupling of transport and payload protocols allow a significant improvement in monitoring signals that are transported through several administrative domains, e.g. a meshed NREN topology where the signals are transported through several other NRENs before reaching the end users.

Why OTN

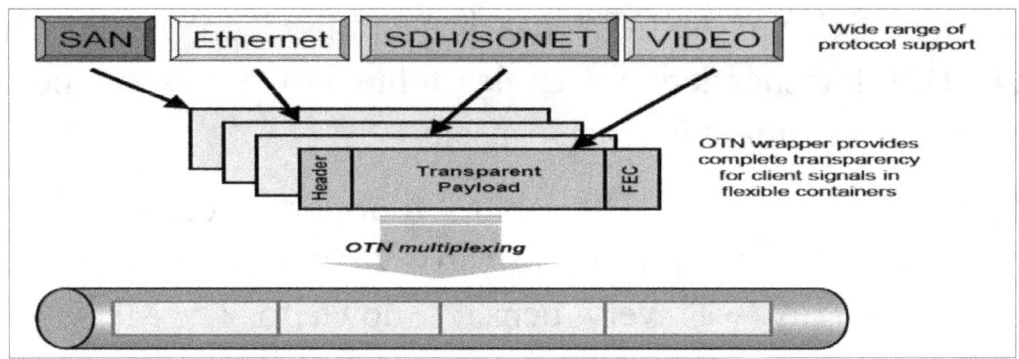
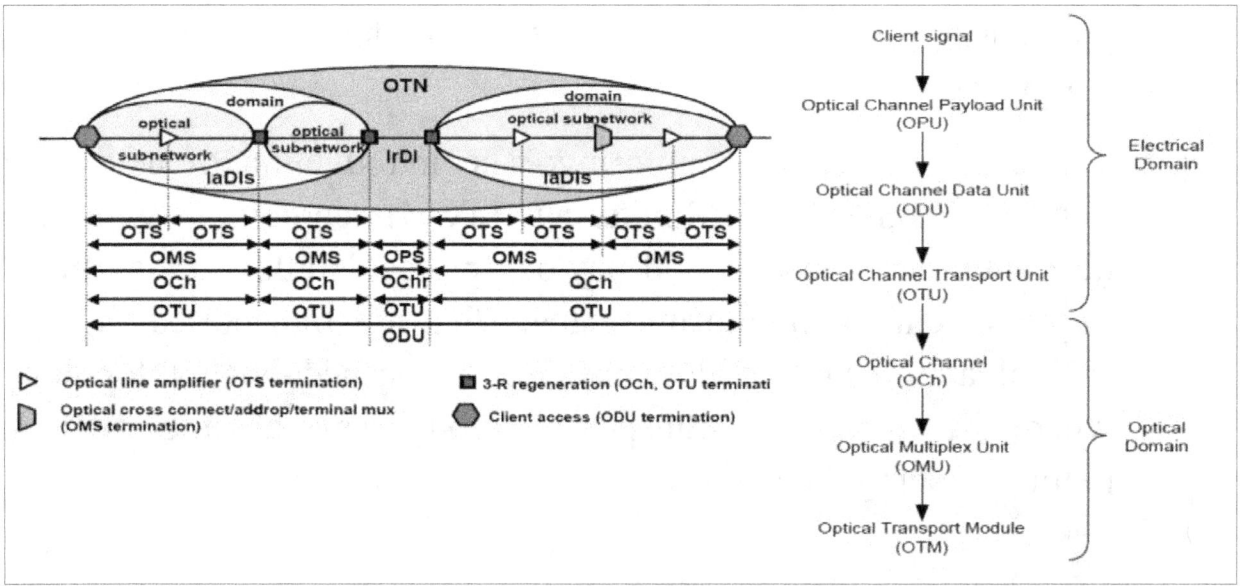
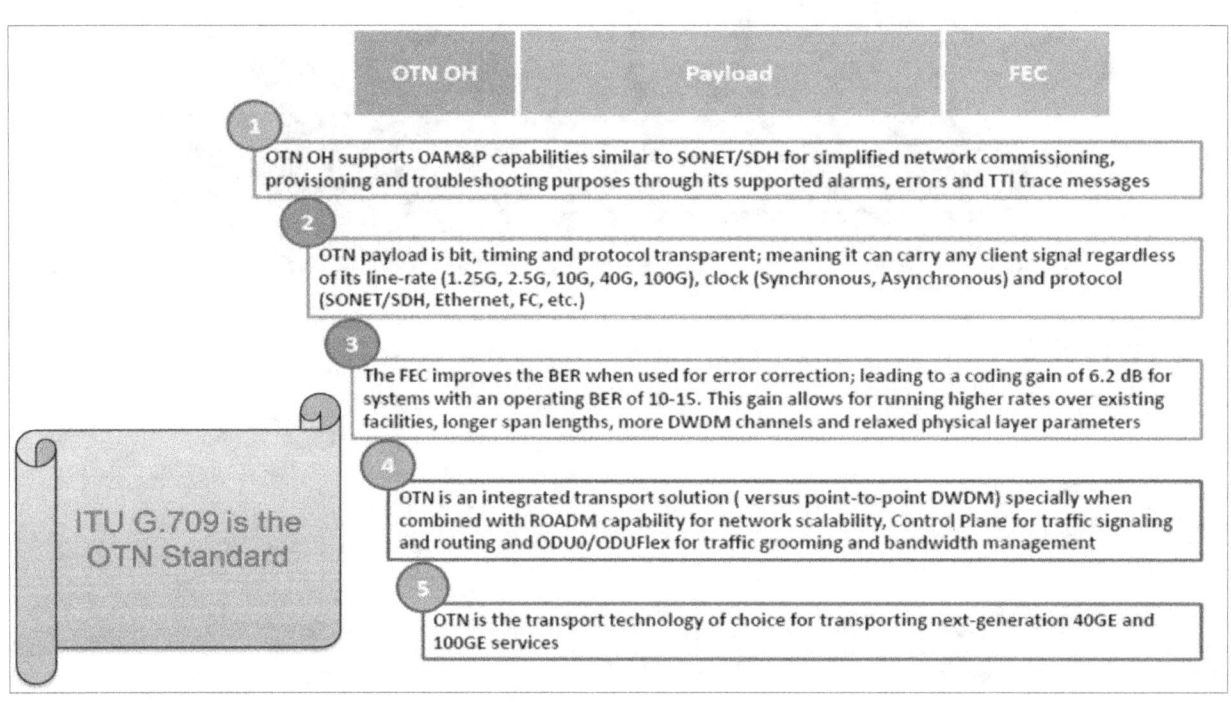

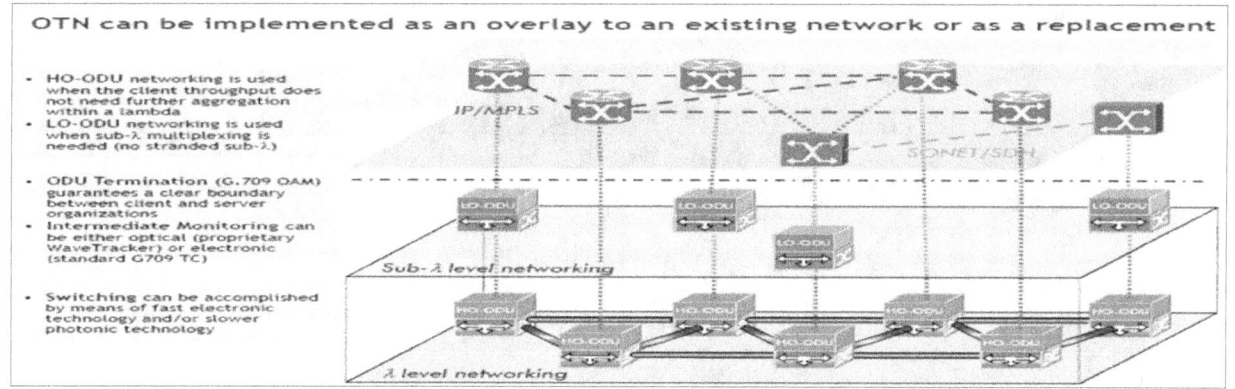

OTN mapping structure

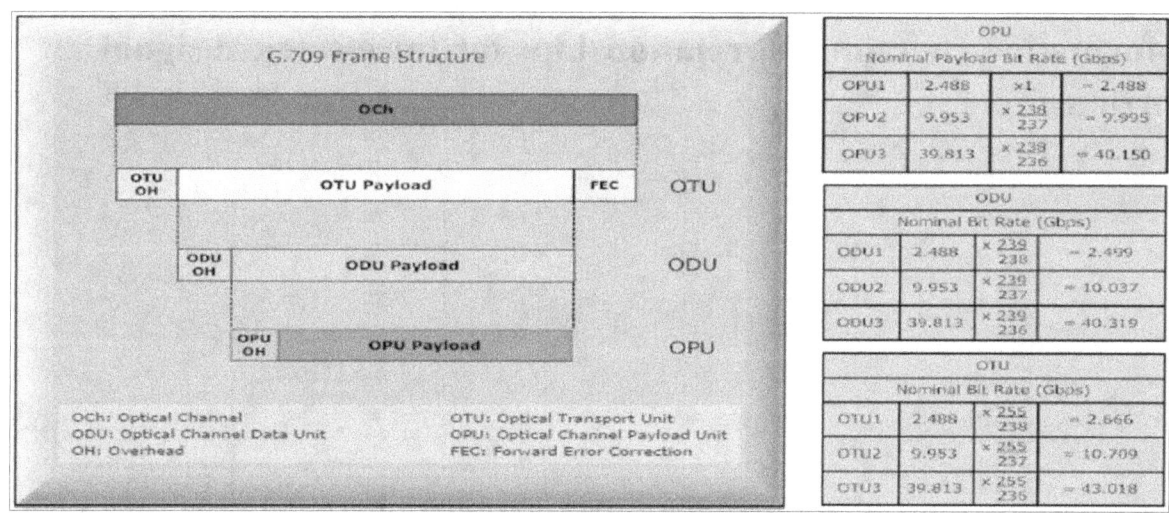

OTN Frame Format

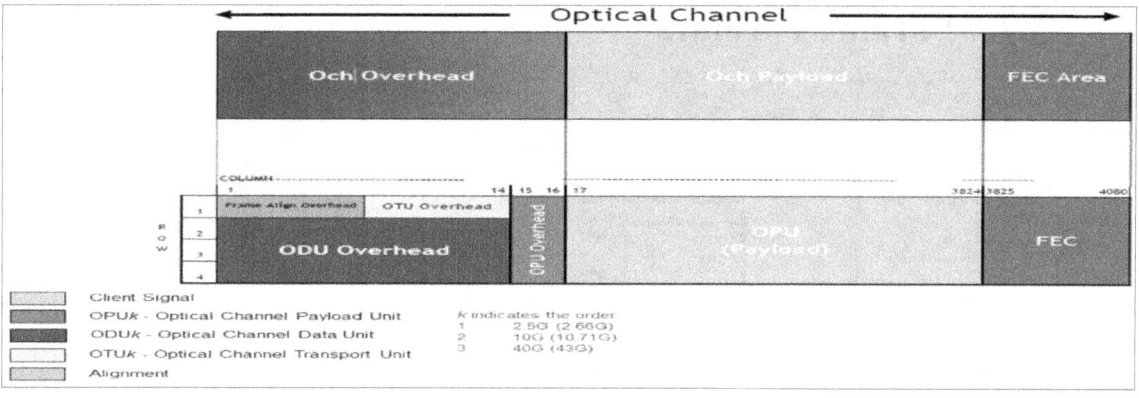

OTN Overhead structure

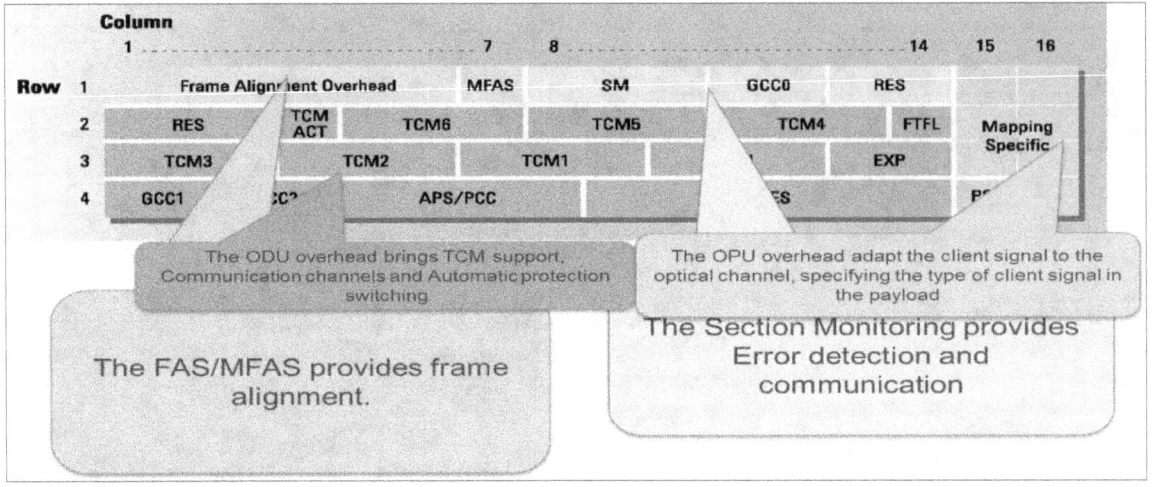

Information containment relationships for the electrical signal portions:

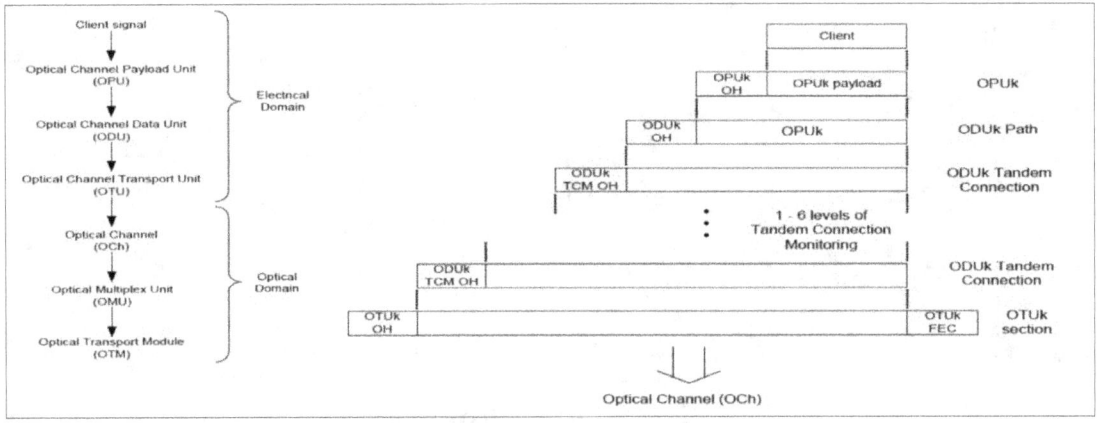

OTN (G.709) Frame Format:

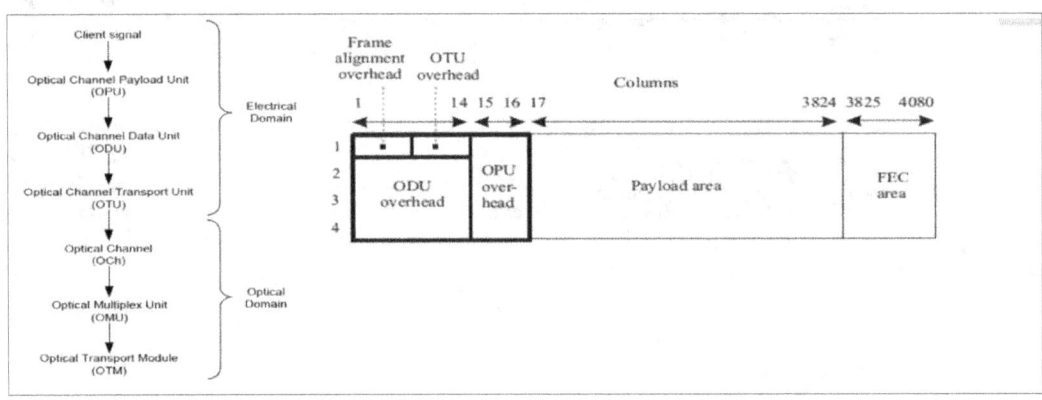

Signal Rates:

k	OTUk signal rate	OPUk payload area rate	OTUk/ODUk/OPUk frame period
0	Not applicable	238/239 × 1 244 160 kbit/s = 1 238 954 kbit/s	98.354 μs
1	255/238 × 2 488 320 kbit/s = 2 666 057 kbit/s	2 488 320 kbit/s	48.971 μs
2	255/237 × 9 953 280 kbit/s = 10 709 225 kbit/s	238/237 × 9 953 280 kbit/s = 9 995 277 kbit/s	12.191 μs
3	255/236 × 39 813 120 kbit/s = 43 018 414 kbit/s	238/236 × 39 813 120 kbit/s = 40 150 519 kbit/s	3.035 μs
4	255/227 × 99 532 800 kbit/s = 111 809 974 kbit/s	238/227 × 99 532 800 kbit/s = 104 355 975 kbit/s	1.168 μs

Note: All rates are ±20 ppm.

4.3 Wave Division Multiplexing (WDM):

As we know, signal transmission through fiber-optic systems becomes increasingly difficult as the data rate on an optical channel increase. Dispersion effects become more significant at higher speeds, and can limit transmission distances, depending on the type of fiber. The faster the channel rate, the shorter the distance signals can travel. Thus 2.5Gbps signal can go farther than 10Gbps signals, and 40Gbps signals cannot go as far as 10Gbps. Thus, one way to achieve higher overall transmission rates over the same distance is to break the signal into many parallel optical channels and transmit them at different wavelengths through the same fiber.

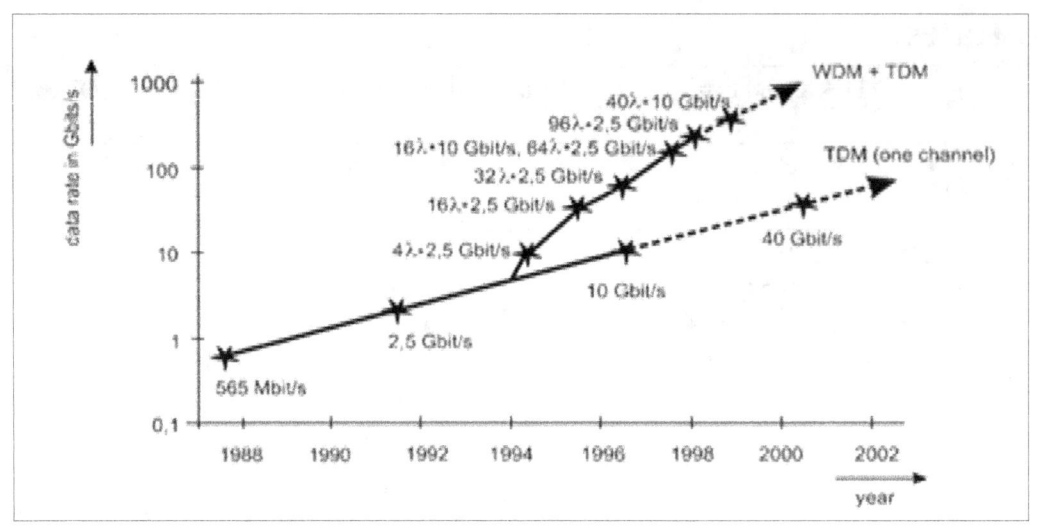

- Cost: Reduce the cost of network and save the fiber, WDM uses one fiber instead of many fibers used by some single-channel systems. WDM uses OA instead of REG. as we know, REG equipment's are expensive, different signal format and bit rate have to use different REG.
- C- High Bandwidth Demand:
- Bandwidth are doubling every 3 months
- Internet traffic increases thousand-fold every 3 years
- upgrade of existing fiber networks (without adding fibers)

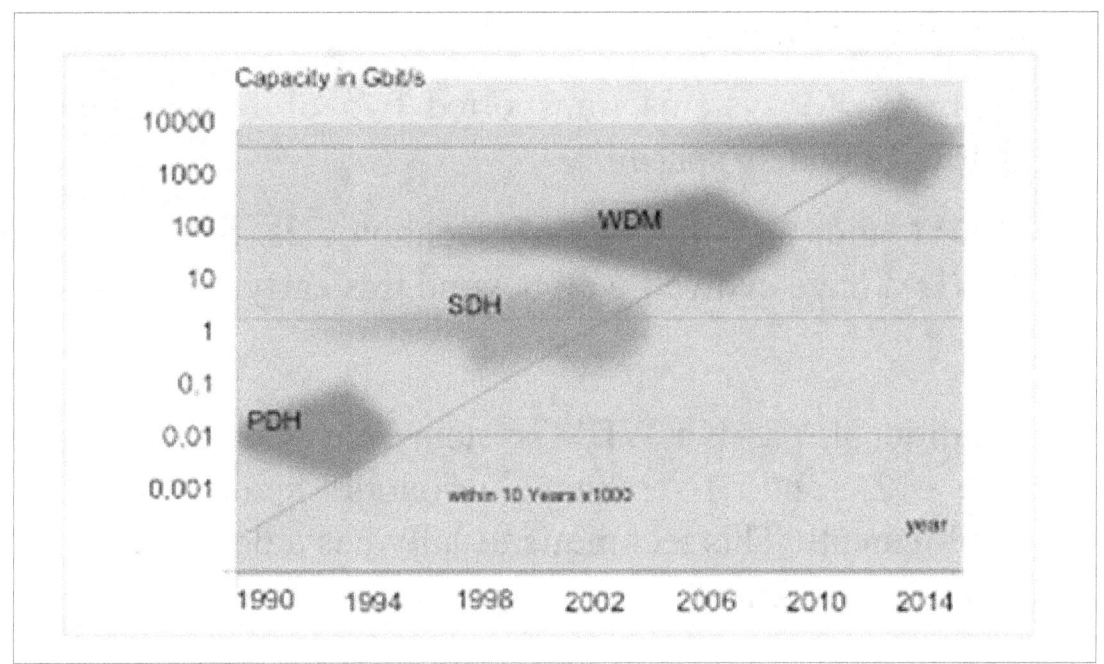

- Transparency: Each optical channel can carry any transmission format (different asynchronous bit rates, analog or digital)
- Scalability: Buy and install equipment for additional demand as needed

What is WDM: WDM (Wavelength-division Multiplexing): WDM is a fundamental passive optical component for optical system. Sending several signals through one WDM with different wavelengths of light. By its special physical character, for instance, a fused dual-channel WDM will be able to double the data transmission bandwidth at very low cost.

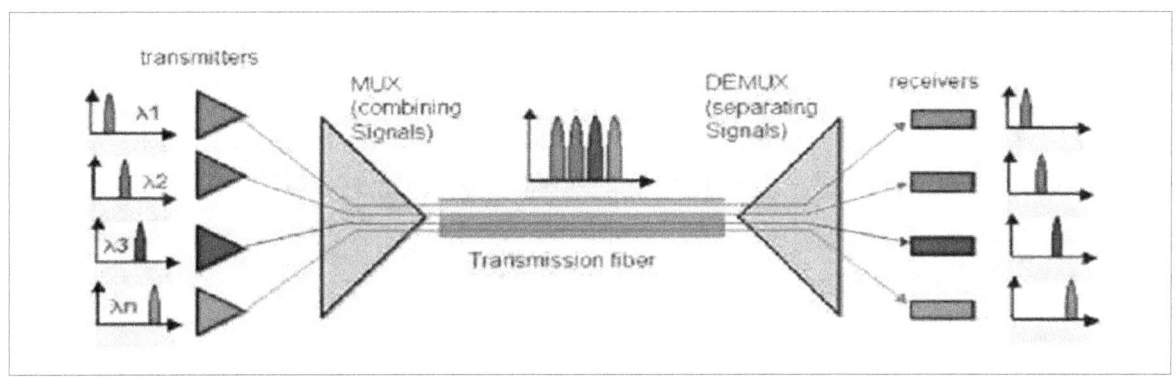

- Varieties of WDM
  - Early WDM systems transported two or four wavelengths that were widely spaced.
  - WDM and the "follow-on" technologies of CWDM and DWDM have evolved well beyond this early limitation.
- WDM
  - Traditional, passive WDM systems are wide-spread with 2, 4, 8, 12, and 16 channel counts being the normal deployments. This technique usually has a distance limitation of under 100 km.
- CWDM: Today, coarse WDM (CWDM) typically uses 20-nm spacing (3000 GHz) of up to 18 channels. The CWDM Recommendation ITU-T G.694.2 provides a grid of wavelengths for target distances up to about 50 km on single mode fibers as specified in ITU-T Recommendations G.652, G.653 and G.655. The CWDM grid is made up of 18 wavelengths defined within the range 1270 nm to 1610 nm spaced by 20 nm.

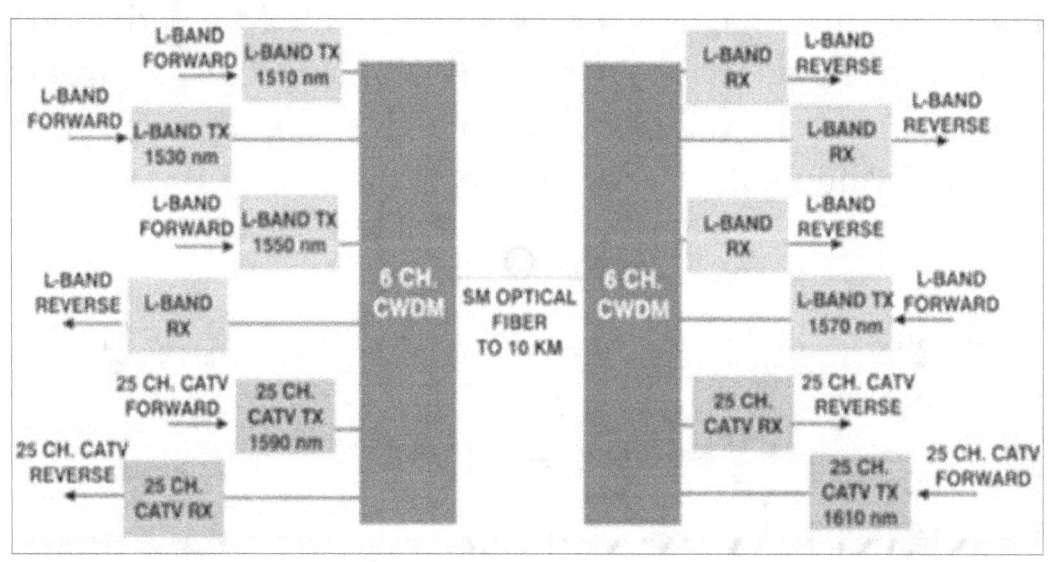

- DWDM: Dense WDM common spacing may be 200, 100, 50, or 25 GHz with channel count reaching up to 128 or more channels at distances of several thousand kilometers with amplification and regeneration along such a route.

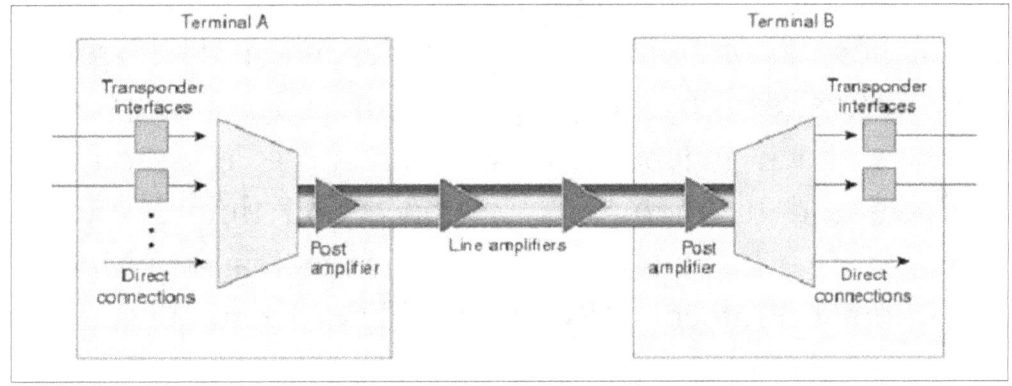

- **WDM Wavelengths features:**

1. We used C_band (1530: 1565 nm) because it faced minimum attenuation.

2. We divided this band into 40 channels with separation 100G HZ for WDM.

3. We divided this band into 80 channels with separation 50G HZ for DWDM.

4. 50G HZ is the minimum separation without interference, and if we want to increase channels, we must use another band (L _band).

- Types of Multiplexing
    - Multiplexing is sending multiple signals or streams of information through a circuit at the same time in the form of a single, complex signal and then recovering the separate signals at the receiving end. Basic types of multiplexing include frequency division (FDM), time division (TDM), and wavelength division (WDM), with TDM and WDM being

widely utilized by telephone and data service providers over optical circuits.

- Time Division Multiplexing

Time-division multiplexing (TDM), as represented in next Figure is a method of combining multiple independent data streams into a single data stream by merging the signals according to a defined sequence. Each independent data stream is reassembled at the receiving end based on the sequence and timing. Synchronous Optical Network (SONET), Asynchronous Transfer Mode (ATM) and Internet Protocol (IP) utilize TDM techniques. In modern telecommunications networks, TDM signals are converted from electrical to optical signals by the SONET network element, for transport over optical fiber.

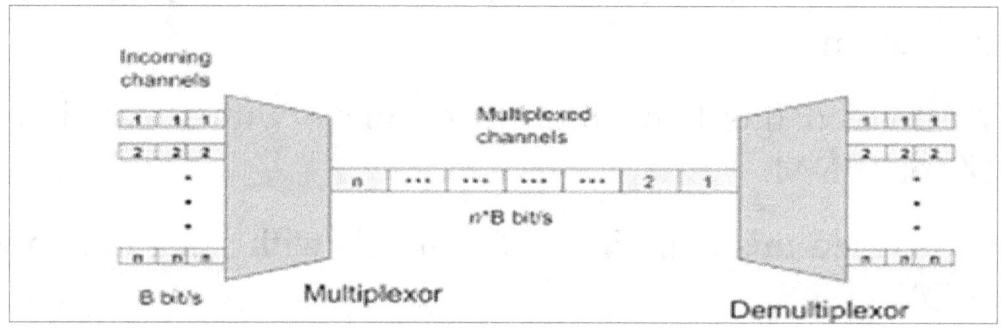

Wavelength Division Multiplexing: WDM combines multiple optical TDM data streams onto one fiber through the use of multiple wavelengths of light. Each individual TDM data stream is sent over an individual laser transmitting a unique wavelength of light.

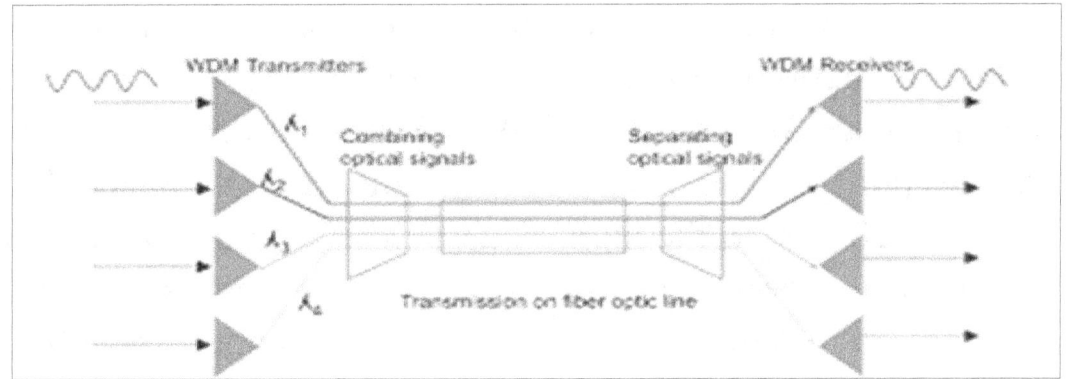

- FDM: Frequency Division Multiplexing (FDM) is a networking technique in which multiple data signals are combined for simultaneous transmission via a shared communication medium. FDM uses a carrier signal at a discrete frequency for each data stream and then combines many modulated signals. When FDM is used to allow multiple users to share a single physical communications medium (i.e. not broadcast through the air), the technology is called frequency-division multiple access (FDMA).

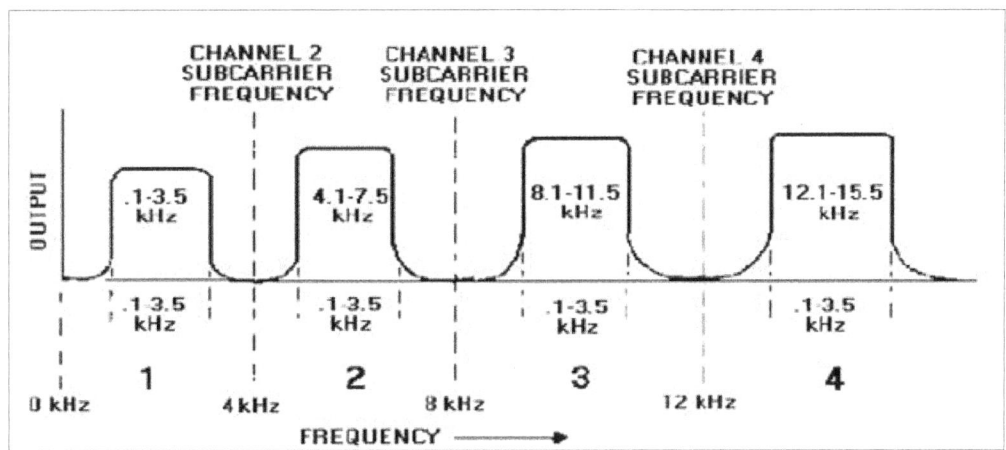

- Transmission Windows: If one looks at the possible wavelengths for the transmission of signals one has to look at the fiber properties. Optical fibers are not suitable for transmission at all wavelengths but only in certain windows. Today, usually the second transmission window (around 1300nm) and the third and fourth transmission windows from

1530 to 1565nm (also called Conventional Band) and from 1565 to1620nm (also called Long Band). are used. Technological reasons limit DWDM applications at the moment to the third and fourth window. The losses caused by the physical effects on the signal due by the type of materials used to produce fibers limit the usable wavelengths to between 1280nm and 1650nm. Within this usable range the techniques used to produce the fibers can cause particular wavelengths to have more loss so we avoid the use of these wavelengths as well

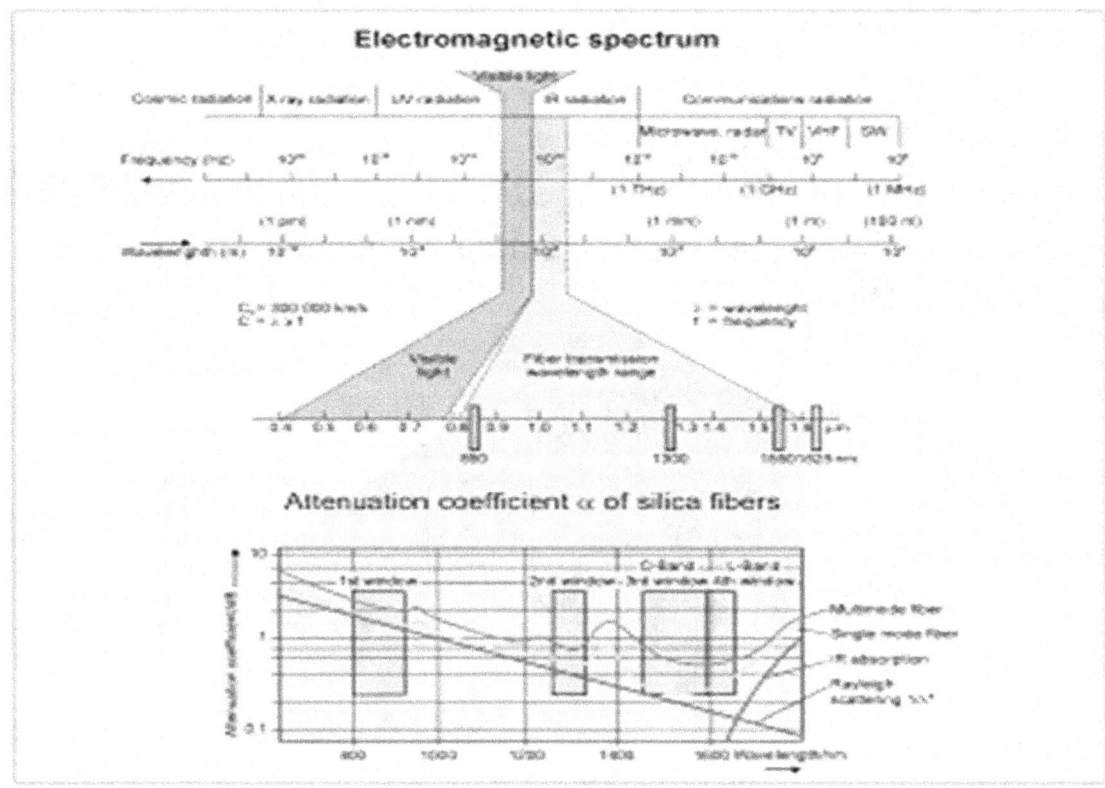

ITU Wavelength Plan:

Within those windows the ITU has defined in G.692 a wavelength plan for DWDM systems to use. In fact, not the wavelengths have been defined, but the frequencies. This does not matter though as the frequency f and the wavelength λ are connected by the relation:

$$C = f * \lambda$$

where c is the speed of light.

These defined frequencies are given by the equation:

$$f = 193.1 \pm m * 0.05 \text{THz}$$

Which means that the ITU in G.692 initially uses a 50GHz grid. There are also proposals for 100GHz and 200GHz spacing or even for unequal channel spacing for specific applications. The wider spacing is easier to handle, but some of the systems existing or planned are in fact already using a 50GHz spacing or even below. Further on, additionally to the C-Band the L-Band will be used in the future.

Optical Supervisory Channel:

In addition to those "working" wavelengths another set of wavelengths (1510 or alternatively 1480 and 1310nm) is reserved for use as the optical supervisory channel, an additional optical channel connecting the DWDM network elements together and fulfilling approximately the same purpose as the SDH overheads.

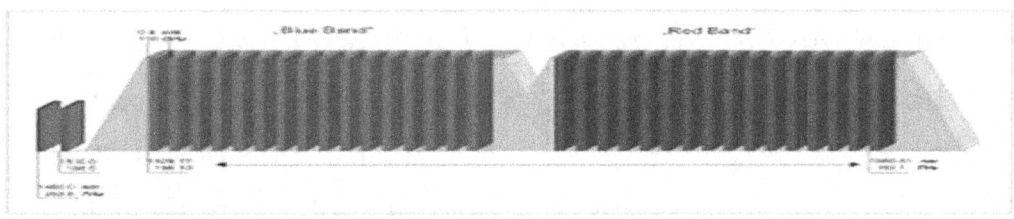

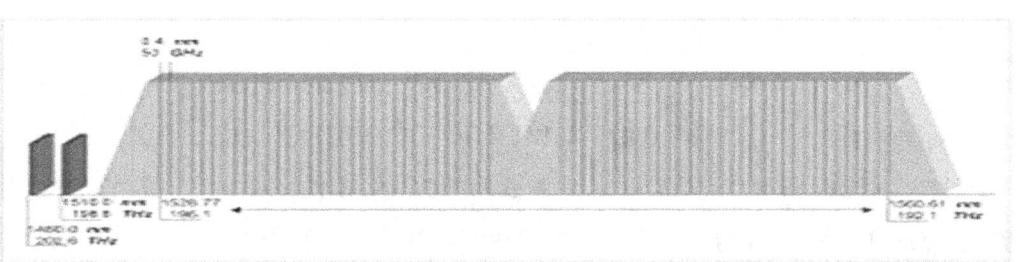

Bandwidth and wavelength

There is a correlation between the frequency f, the propagation velocity v („phase velocity ") and the wavelength λ. In this the frequency f is determined by the processes during the generation of radiation. The medium, in which the wave is propagating, determines the phase velocity v. Consequently, the wave-length λ is no independent quantity. It results from the frequency and the phase velocity. Thus, the light has the same frequency but different wavelengths indifferent substrates. For the propagation in a vacuum it is:

$$C * f = \lambda$$

In this, c is the vacuum velocity of light and λ the wavelength in the vacuum. All correlations stated in the following between frequency and wave lengths refer to the wavelengths in the vacuum.

In principle, with the standard single mode fiber for telecommunication a wavelength range of approx. λ1=1280nm to λ2=1650nm can be utilized. In this, the lower wavelength limit results from the core diameter of the single mode fiber. The upper wavelength limit results from the fact that above this limit the attenuation coefficient rapidly increases and the fiber gets very sensitive regarding macro bending. Corresponding to the pervious equation the resulting usable wavelength range is from f1=235THz to f2=182THz. In this, THz=Terahertz oscillations per second. Thus, the intrinsic transmission capacity of the single mode fiber is:

$$\text{Intrinsic } f \text{ THz} = 1253 \text{ transmission capacity}$$

This transmission capacity is often called „bandwidth of the fiber ". From the equation it is obvious that the transmission capacity of the single mode fiber is only used at a very small scale at present. A 2,5Gbit/s signal, for example, only uses this bandwidth capacity with 0,005% and a 10Gbit/s signal with 0,02%! It is obvious that the transmission capacity of a single mode fiber can be exploited much better by a simultaneous transmission of several.

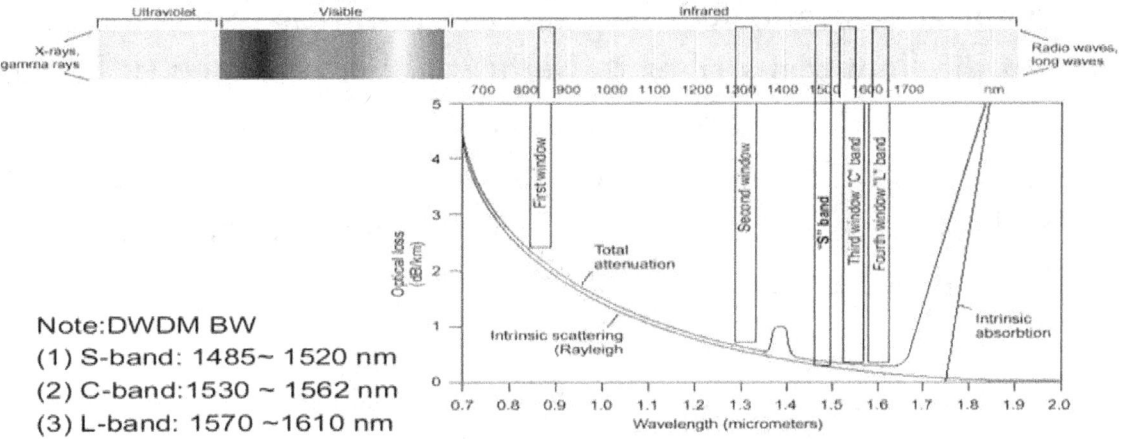

Note: DWDM BW
(1) S-band: 1485~ 1520 nm
(2) C-band: 1530 ~ 1562 nm
(3) L-band: 1570 ~ 1610 nm

Frequency/ THz	Center wavelength/nm	Frequency/ THz	Center wavelength/nm	Frequency/ THz	Center wavelength/nm
195,9	1530,33	194,4	1542,14	192,9	1554,13
195,8	1531,12	194,3	1542,94	192,8	1554,94
195,7	1531,90	194,2	1543,73	192,7	1555,75
195,6	1532,68	194,1	1544,53	192,6	1556,55
195,5	1533,47	194,0	1545,32	192,5	1557,36
195,4	1534,25	193,9	1546,12	192,4	1558,17
195,3	1535,04	193,8	1546,92	192,3	1558,98
195,2	1535,82	193,7	1547,72	192,2	1559,79
195,1	1536,61	193,6	1548,51	192,1	1560,61
195,0	1537,40	193,5	1549,32	192,0	1561,42
194,9	1538,19	193,4	1550,12	191,9	1562,23
194,8	1538,98	193,3	1550,92	191,8	1563,05
194,7	1539,77	193,2	1551,72	191,7	1563,86
194,6	1540,56	193,1	1552,52		
194,5	1541,35	193,0	1553,33		

DWDM wavelength according to the ITU recommendation

WDM parameters:

I. Optical Fibers

The main job of optical fibers is to guide light waves with a minimum of attenuation (loss of signal). Optical fibers are composed of fine threads of glass in layers, called the core and cladding that can transmit light at about two-thirds the speed of light in a vacuum. Though admittedly an oversimplification, the transmission of light in optical fiber is commonly explained using the principle of total internal reflection. With this phenomenon, 100 percent of light that strikes a surface is reflected. By contrast, a mirror reflects about 90 percent of the light that strikes it. Light is either reflected (it bounces back) or refracted (its angle is altered while passing through a different medium) depending upon the angle of incidence (the angle at which light strikes the interface between an optically denser and optically thinner material).

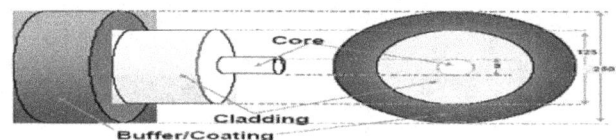

Total internal reflection happens when the following conditions are met:

Beams pass from a denser to a less dense material. The difference between the optical density of a given material and a vacuum is the material's refractive index. Depending upon the angle of incidence (the angle at which light strikes the interface between an optically denser and optically thinner material). Total internal reflection happens when the following conditions are met:

- Beams pass from a denser to a less dense material

- The core has a higher refractive index than the cladding, allowing the beam that strikes that surface at less than the critical angle to

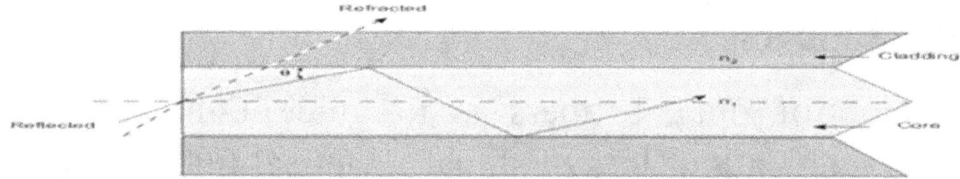

- An optical fiber consists of two different types of highly pure, solid glass (silica) the core and the Cladding that are mixed with specific elements, called dopants, to adjust their refractive indices. The Difference between the refractive indices of the two materials causes most of the transmitted light to bounce off the cladding and stay within the core. The critical angle requirement is met by controlling the angle at which the light is injected into the fiber. Two or more layers of protective coating around the cladding ensure that the glass can be handled without damage.

Types of fibers:

There are two general categories of optical fiber in use today, multimode fiber and single-mode fiber.

a- Multimode fiber:

- Core diameter varies 50 μm for step index 62.5 μm for graded index
- Bit rate-distance product> 500 MHz-k
- Distance limited

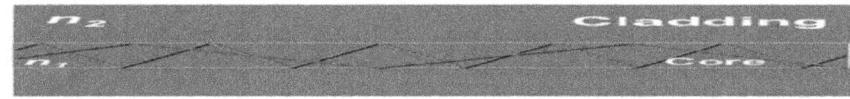

b- single mode fiber:

- Core diameter is about 9 μm
- Bit rate-distance product > 100 THz-km

Designs of single-mode fiber have evolved over several decades.

The three principle types and their ITU-T specifications are:

• Non-dispersion-shifted fiber (NDSF), G.652

• Dispersion-shifted fiber (DSF), G.653

• Non-zero dispersion-shifted fiber (NZ-DSF), G.655

SMF (G.652)	Good for TDM at 1310 nm OK for TDM at 1550 OK for DWDM (with Dispersion Mgmt)
DSF (G.653)	OK for TDM at 1310 nm Good for TDM at 1550 nm Bad for DWDM (C-Band)
NZDSF (G.655)	OK for TDM at 1310 nm Good for TDM at 1550 nm Good for DWDM (C + L Bands)
Extended Band (G.652.C) (Suppressed Attenuation in the Traditional Water Peak Region)	Good for TDM at 1310 nm OK for TDM at 1550 nm OK for DWDM (with Dispersion Mgmt Good for CWDM (> Eight wavelengths)

Limitation of transmission in Optical Fibers:

The optical transmission through optical fiber causes several effects such:

a. Linear Effects: can be compensated:

- Attenuation
- Dispersion
- Non-Linear Effects: will accumulate (not so critical in short-haul network)

b. Polarization Mode Dispersion (not a problem at speeds < OC-192)
- Stimulated Raman Scattering
- Stimulated Brillouin Scattering
- Self-Phase Modulation
- Four-Wave Mixing (The most critical effect; will limit the channel capacity of DWDM system)

a- Linear Effects

A-1 attenuation

- Attenuation in Fiber is caused by two things:

- Absorption by the fiber material
- Scattering of the light from the fiber

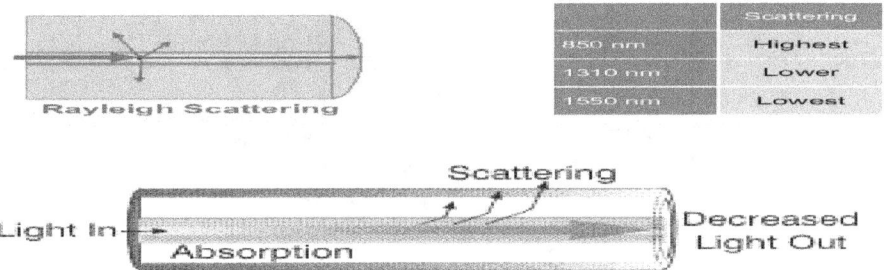

Other Causes of Attenuation in Fiber:

- Micro bends—Caused by small distortions of the fiber in manufacturing
- Macro bends—Caused by wrapping fiber around a corner with too small a bending radius
- Back reflections—Caused by reflections at fiber ends, like connectors
- Fiber splices—Caused by poor alignment or dirt
- Mechanical connections—Physical gaps between fibers

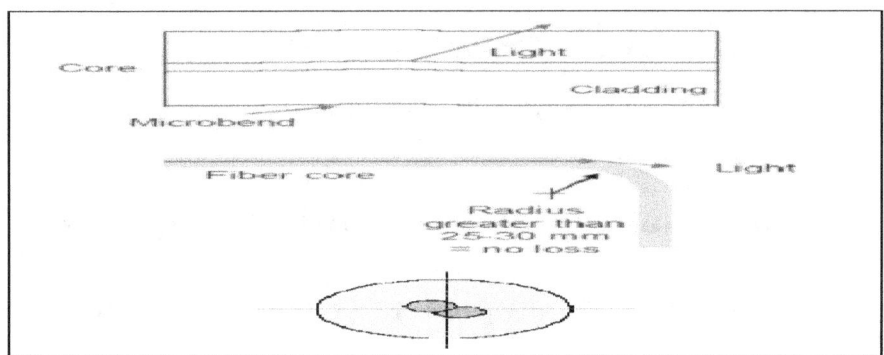

Optical Attenuation

Pulse amplitude reduction limits "how far" (distance)

Attenuation in dB=10xLog (Pi/Po)

Power is measured in dBm: P(dBm)=10xlog (P mW/1 mW

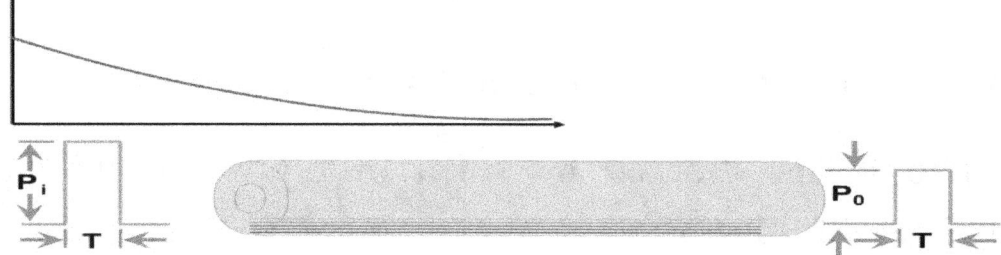

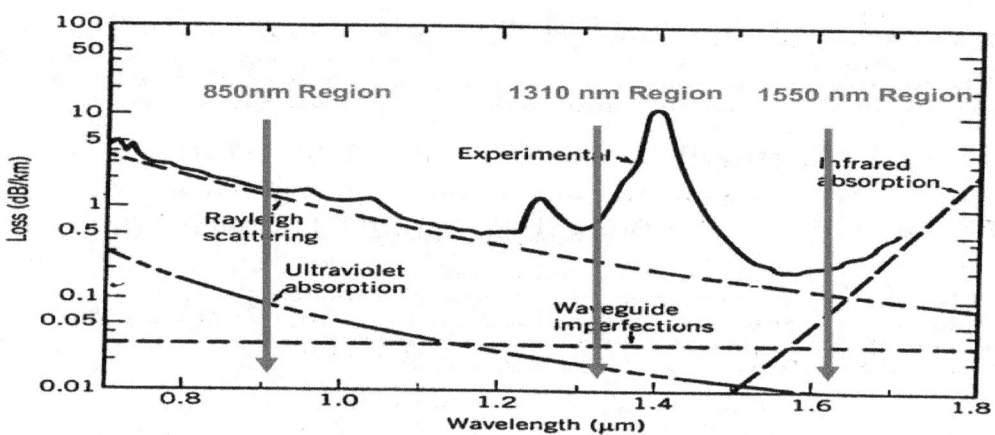

Attenuation Response at Different Wavelengths:

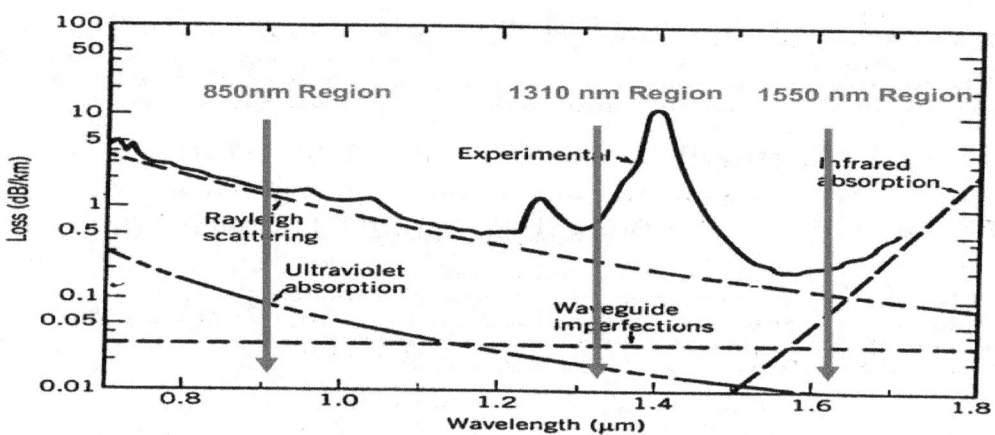

Attenuation: Compensated by Optical Amplifiers

- Erbium-doped fiber amplifiers (EDFA) are the most commonly deployed optical amplifiers. Commercially available since the early 1990s Works best in the range 1530 to 1565 nm Gain up to 30 dB (1000 photons out per one photon in):
  - ➤ Optically transparent

- Wavelength transparent
- Bit rate transparent

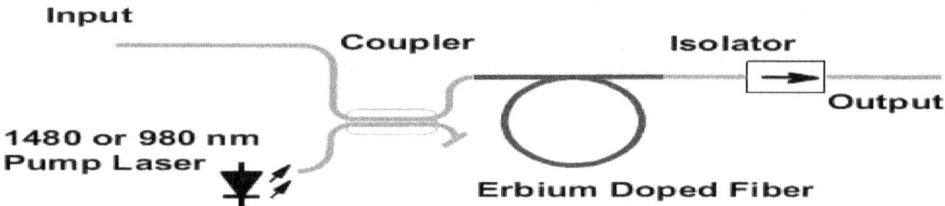

A-2 Dispersion

Dispersion is based primarily on the fact that the refractive index of glass n (the optical "density") depends on the wavelength of light, which in turn causes these different wavelengths to travel at different speeds in the same medium (chromatic dispersion). In transmission technology, dispersion therefore is the tendency of optical pulses to spread as they travel through the optical fiber. As a consequence, it becomes more difficult to distinguish if a received bit is '1' or '0'. This effect is called Inter-Symbol-Interference (ISI). The problem of dispersion becomes critical on long fibers carrying high bandwidth signals. Total dispersion is measured in units of ps/nm and the dispersion coefficient of a fiber is in units of ps/nm-km, or pico-seconds per nanometer per km of fiber length.

Mathematically speaking the dispersion is the differential of the Index of Refraction n to the wavelength λ.

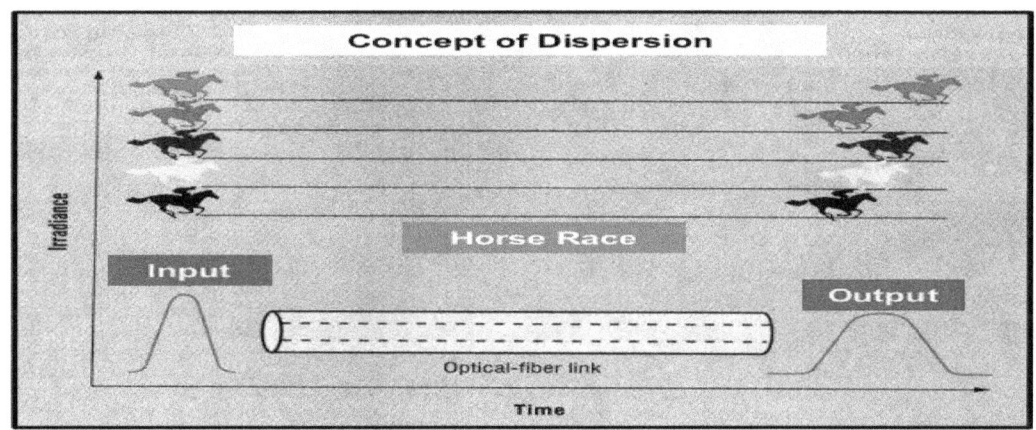

In fact, the phenomenon "Dispersion" is composed of several factors:

- Mode-Dispersion:

Different modes in a multimode fiber have different paths and therefore different travelling times (not relevant in single mode fibers).

- Profile Dispersion:

For graded index fibers impurities or imperfections in the grading profile cause additional dispersion.

- Chromatic Dispersion

In the mono-mode fiber the chromatic dispersion is the dominant factor. It is composed from:

- Material Dispersion

This is a pure physical property and describes the different speed of different wavelengths in the glass used. It can be changed only very little by the doping of the fiber.

- Waveguide Dispersion

This type of dispersion is highly dependent on the profile of the graded index. Thus, it is possible to influence the total dispersion of the fiber.

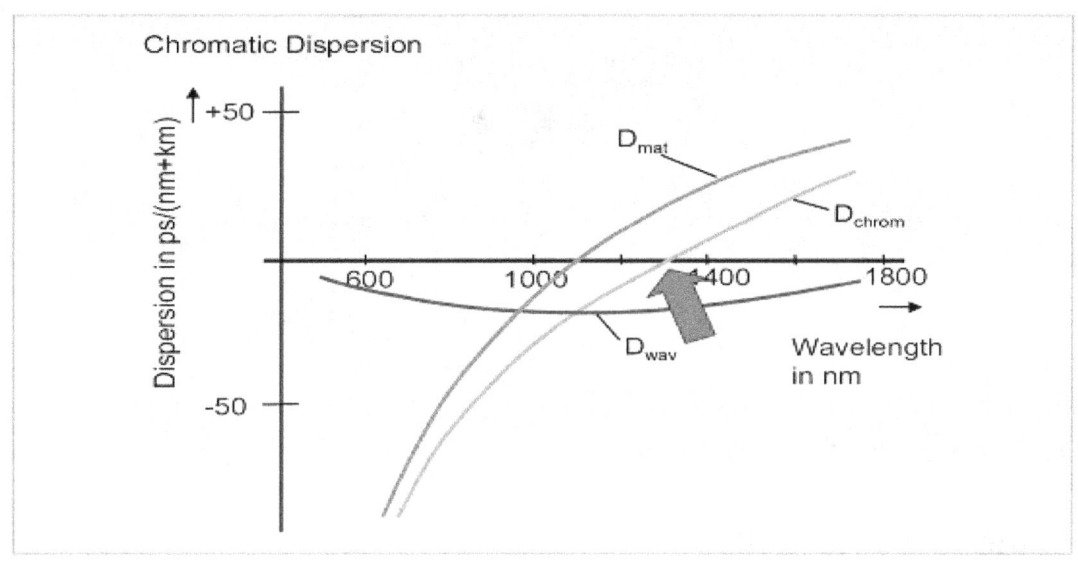

B-Non linear effects:

B-1 Polarization Mode Dispersion

The base mode in mono-mode fibers consists of two orthogonally polarized Parts or polarization modes. Due to environmental conditions (stretching, bending, torsion...) the radial symmetry of the index profile is disturbed. Consequently, those two parts have different speeds, leading to a temporal broadening of the pulse. The result of PMD is additional ISI. Due to a coupling of those polarization modes, the total delay and pulse broadening is only dependent on the square root of the fiber length: The latter is a fiber property, which is of course depending on a lot of environmental conditions and which is usually specified as an average value by the producer.

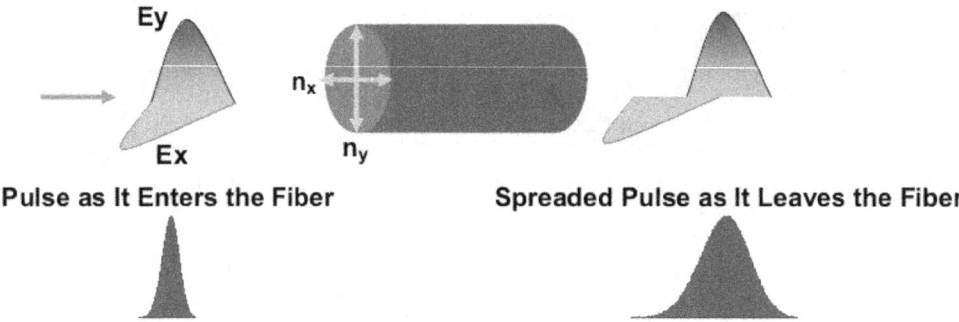

Pulse as It Enters the Fiber — Spreaded Pulse as It Leaves the Fiber

Contributing factors are e.g.:

- Fiber core elasticity
- Transverse stress
- Bending
- Twisting
- Isolators in EDFAs
- Aging

In usual systems the PMD is of no practical relevance. In DWDM systems though, which are dispersion-managed and use extremely narrow bandwidths, the PMD becomes a very considerable factor limiting especially high-bitrate transmission. One of the biggest problems of PMD is, that it might be highly time dependent. This becomes clear, when we imagine e.g. a truck or a train passing by a fiber line, causing vibrations which in turn might cause fiber deformations.

$$\text{Distance (Km)} = \frac{\text{Specification of Transponder (ps/nm)}}{\text{Coefficient of Dispersion of Fiber (ps/nm*km)}}$$

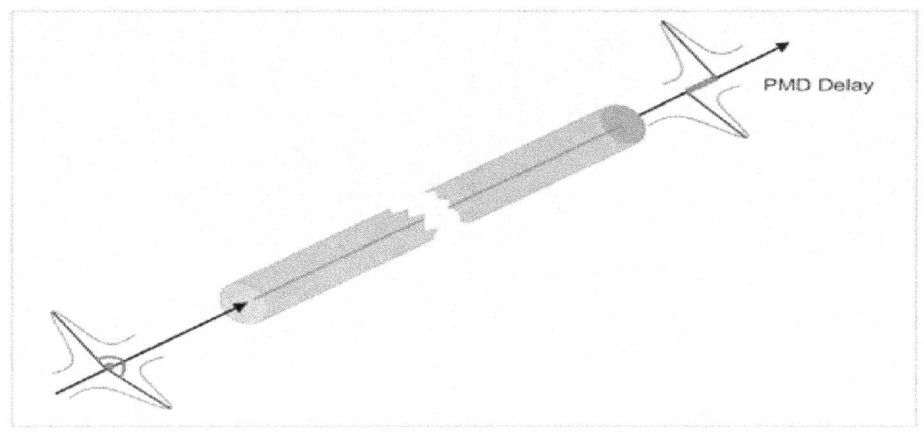

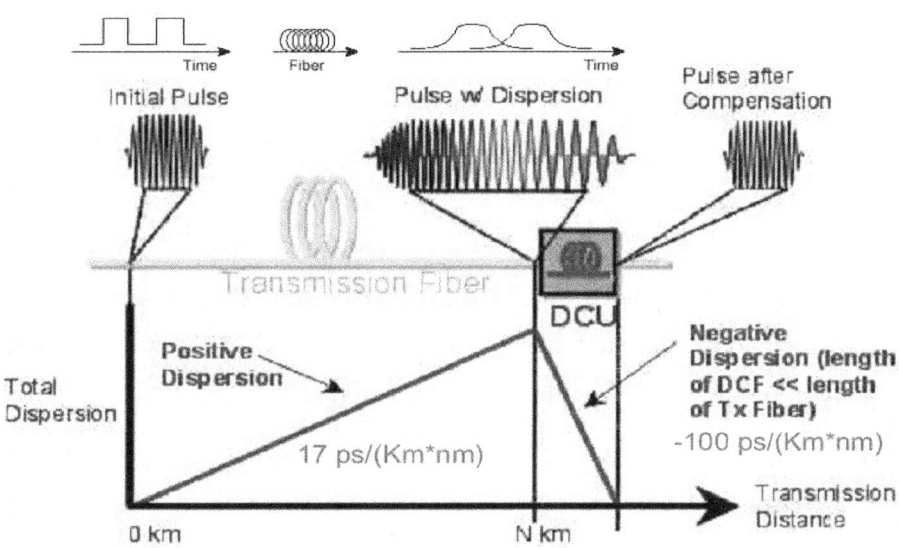

B-2 Stimulated Raman Scattering:

A special form of scattering is the interaction of Photons and optical Phonons. The latter are the particles, which can be assigned to vibrations of the array of atoms and electrons. Like in playing Billiard, a Photon might hit a Phonon and glance of, losing a bit of its energy. Alternatively, a Phonon might hit a Photon, adding a bit to its energy. The result is, that an initial peak after scattering has two peaks on the sides at higher and lower wavelengths. The one at the higher energetic side usually can neglected though, as there are not so many Phonons around that are able to increase the Photons energy.

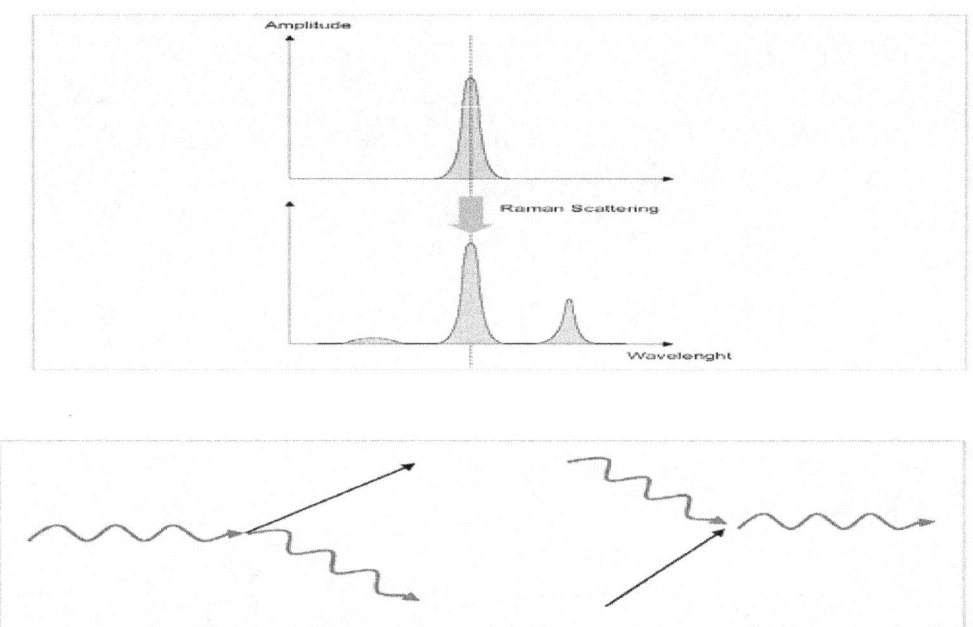

If the Raman Scattering takes place under the influence of bypassing light of a fitting frequency, the scattering produces a photon of exactly the same frequency. The rest of the energy is emitted as a phonon, effectively as vibrations of the atomic structure. This is then called "Stimulated Raman Scattering" In DWDM systems the higher frequency channels lose energy to channels with lower frequencies (Raman-Tilt) resulting in channel crosstalk and reinforcing other nonlinear effects like Cross-Phase modulation. Using pre-emphasis can compensate this effect.

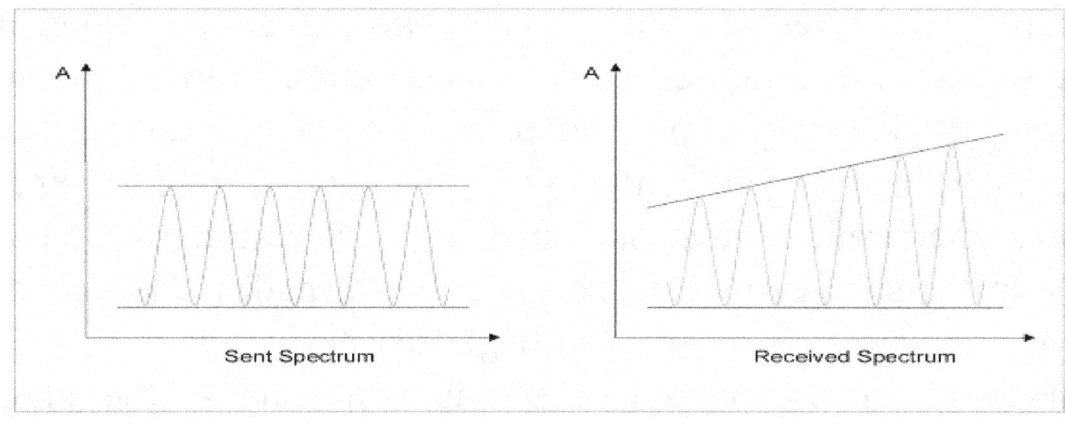

B-3 Stimulated Brillouin Scattering

An effect similar to Raman Scattering is Brillouin Scattering, the difference mainly being that another type of phonon (so-called acoustic phonons) is responsible for the scattering. It also is a source of crosstalk, though the peaks are influenced in a different way to Raman Scattering, the scattered light being downshifted by approximately 11GHz at 1550nm. As SBS acts in backward direction, it is especially important when there is bi-directional. What happens is that a channel interferes with itself causing significant distortion and loss in forward direction. Like the other non-linear effects, SBS occurs only significantly when crossing a certain power threshold. One method of reducing the influence of SBS is to modulate the transmission laser with a very low frequency which is sometimes also called "pilot tone", as SBS gets lower with rising line width.

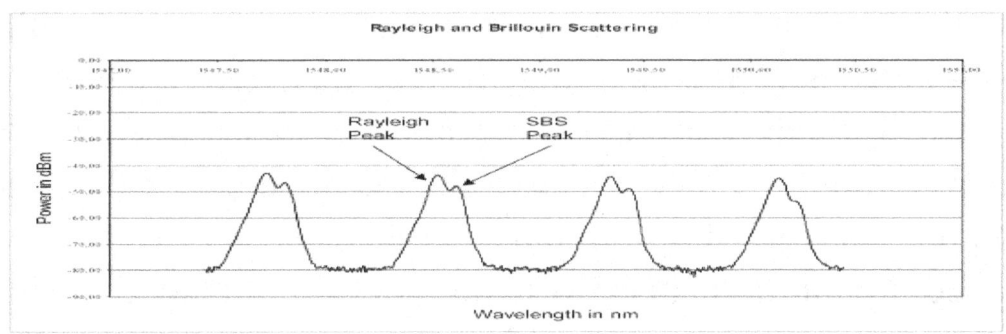

Rayleigh and SBS peak

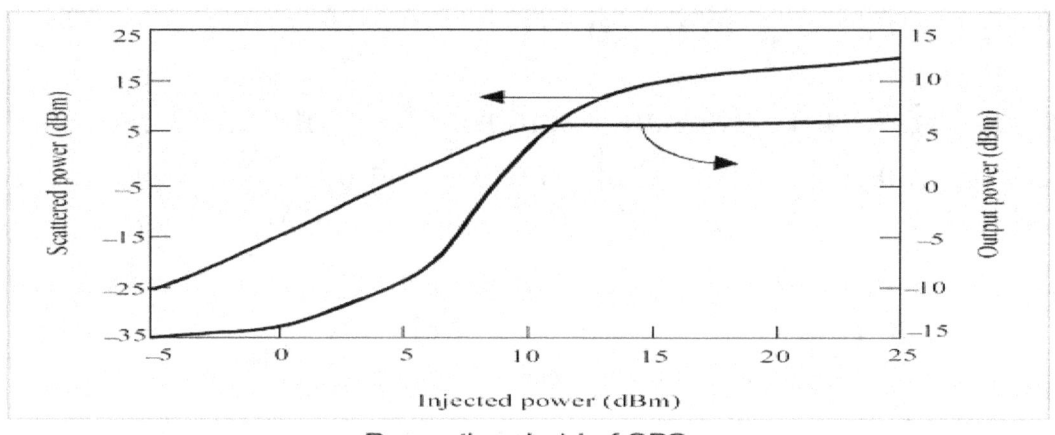

Power threshold of SBS

B-4 Four-wave-mixing

Four-wave-mixing is the first of the Kerr-non-linear effects. It is due to the fact that the index of refraction has a power-dependent component:

$$n = n(\lambda) + nKerr * P.$$

It is the dipole nature of the atoms that leads to interaction with light and to oscillations of the electrons. In that way light changes matter and in turn is changed by the same matter, thus influencing itself. That causes - among other effects – the dependency of the speed of light on the power of an impulse! FWM is an effect of great importance in multi-channel systems. Let's imagine 3 signals at different wavelengths entering a fiber. The phenomenon FWM is an interaction between 4 photons or waves and causes signals to be produced at frequencies

FFWM=fx+fy-fz for all possible combinations of fx,fy,fz.

The frequency f123 is therefore obtained by the equation f123=f1+f2-f3. The total number of these "ghost"-wavelengths is given by: Number of Ghosts = $0.5N^2(N-1)$ where N is the number of channels. For a 16-channel system that means we have 1920 ghosts!

This of course once more means crosstalk, especially in systems with an equally spaced wavelength grid like we have in DWDM systems.

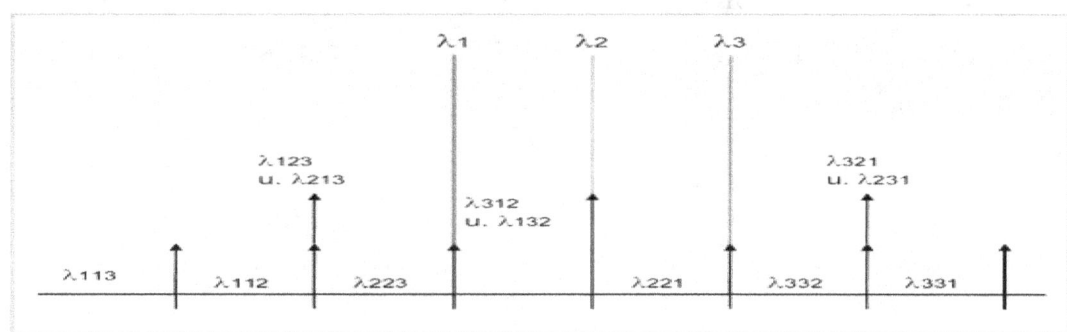

Four wave mixing

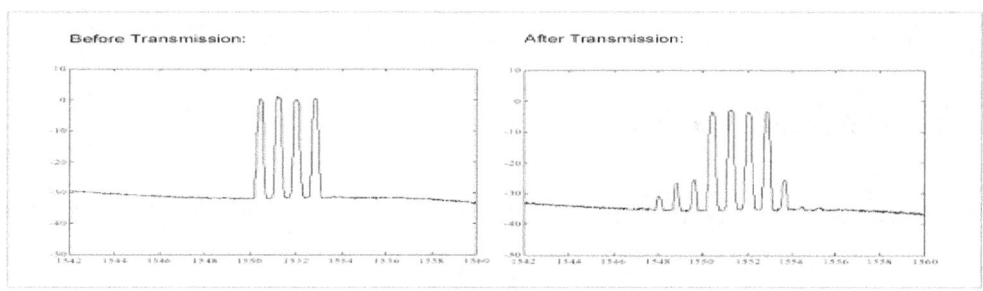

FWM measurement

One way to control FWM is dispersion: the higher the dispersion, the less important the FWM. This happens because with low dispersion all the channels travel in the fiber at the same speed so the light signals are spatially close to each other. This way each of the channels can easily influence the others. With high dispersion the channels travel at different speeds (so called pulse walk-off so they are never spatially close and the effect of FWM is reduced.

Other ways of influencing FWM include e.g. a rise in channel spacing or the use of unequal channel spacing. Self-Phase Modulation SPM is a direct consequence of the refraction index dependence with power, because the speed at which light travels depends on the refraction index. What happens is that some parts of an impulse travel slower than others resulting in a broadening or compression of the impulse. The result is a bit similar to dispersion: the pulse width gets wider. The net result of SPM and dispersion depends on the dispersion regime: Below the zero-dispersion point the pulse gets wider because dispersion and SPM act „in the same direction ". Above that point dispersion and SPM. Compensate for each other, reducing broadening.

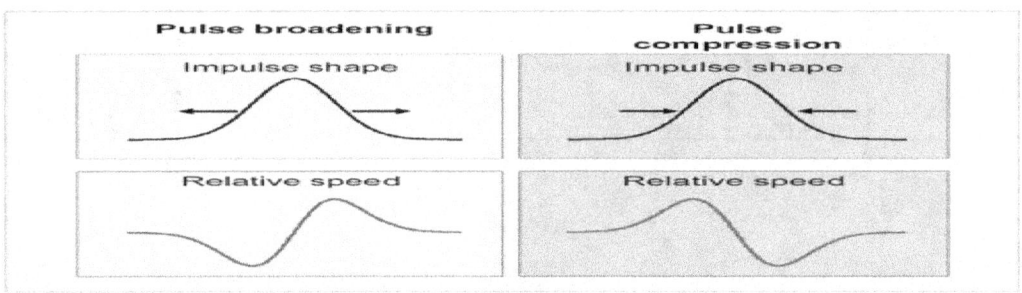

Effects of SPM above and below zero dispersion point

Cross Phase Modulation

The last of the nonlinear effects we are going to mention is the Cross-Phase Modulation (XPM). XPM is basically the same as SPM. The difference is that the refraction index is now influenced by the power of all the channels in the WDM signal. Combined with dispersion XPM is the dominant non-linear effect in SMF, resulting in high crosstalk between channels.

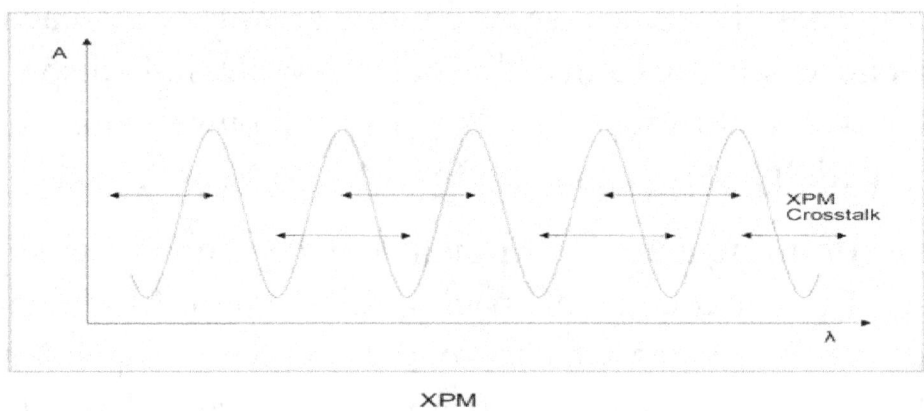

XPM

1- Transponder

Convert from color to (black & white) optical signal. Used transponder terminal that convert from black & white to color signal from black & white equipments. There are some equipments deal with color signal directly without transponder

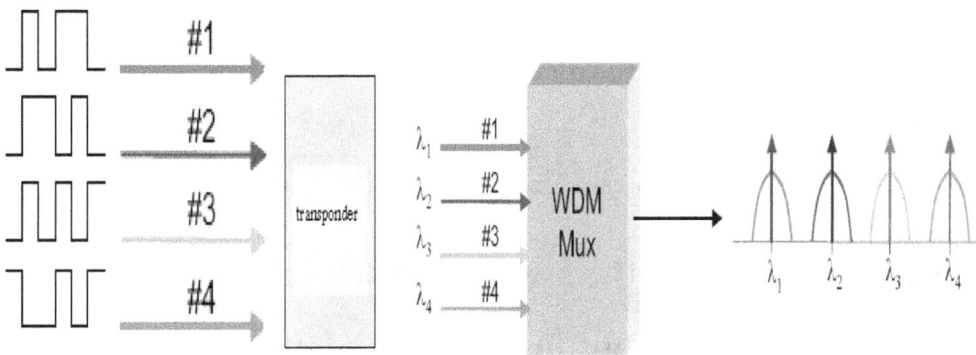

Within the DWDM system a transponder converts the client optical signal from back to an electrical signal and performs the 3R functions (see Figure). This electrical signal is then used to drive the WDM laser. Each transponder within the system converts its client's signal to a slightly differentwavelength. The wavelengths from all of the transponders in the system are then optically multiplexed.

In the receive direction of the DWDM system, the reverse process takes place. Individual wavelengths are filtered from the multiplexed fiber and fed to individual transponders, which convert the signal to electrical and drive a standard interface to the client. Future designs include passive interfaces, which accept the ITU-compliant light directly from anattached switch or router with an optical interface.Operation of a Transponder Based DWDM System the next Figure shows the end-to-end operation of a unidirectional DWDM system.

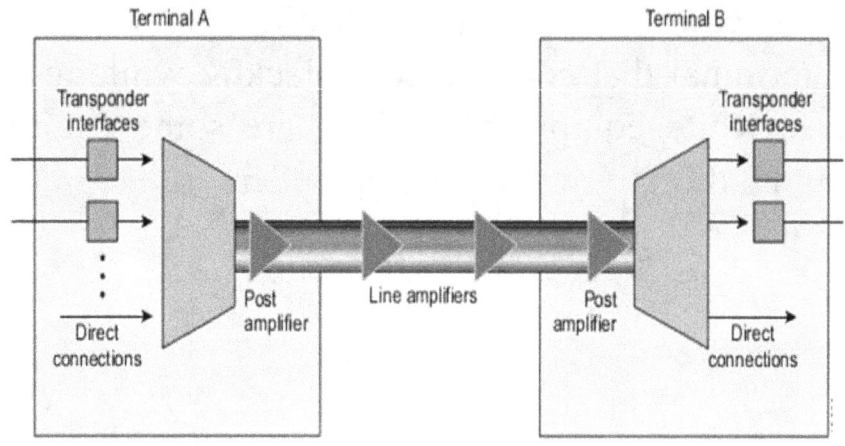

Anatomy of a DWDM System

The following steps describe the system shown in Figure

1. The transponder accepts input in the form of standard single-mode or multimode laser. The input can come from different physical media and different protocols and traffic types.

2. The wavelength of each input signal is mapped to a DWDM wavelength.

3. DWDM wavelengths from the transponder are multiplexed into a single optical signal and launched into the fiber. The system might also include the ability to accept direct optical signals to themultiplexer; such signals could come, for example, from a satellite node.

4. A post-amplifier boosts the strength of the optical signal as it leaves the system (optional).

5. Optical amplifiers are used along the fiber span as needed (optional).

6. A pre-amplifier boosts the signal before it enters the end system (optional).

7. The incoming signal is demultiplexed into individual DWDM lambdas (or wavelengths).

8. The individual DWDM lambdas are mapped to the required output type (for example, OC-48 single-mode fiber) and sent out through the transponder.

2- Filters and Gratings: For DWDM it is essential to have the ability to filter out one particular wavelength. Several filtering methods exist; most of these techniques are in one way or another using interference. A good optical filter for DWDM is characterized by the capacity of isolation (eliminate power from other channels) and distortion (to minimize signal distortion due to filter response). If a filter does not provide good isolation then there will be signal degradation due to linear crosstalk. That means that power from other channels will reach the receiver, interfering with the selected channel. But, in order to achieve good isolation, it is necessary to reduce the filter bandwidth, thus increasing filter distortion. The system designer has to find a compromise between these factors.

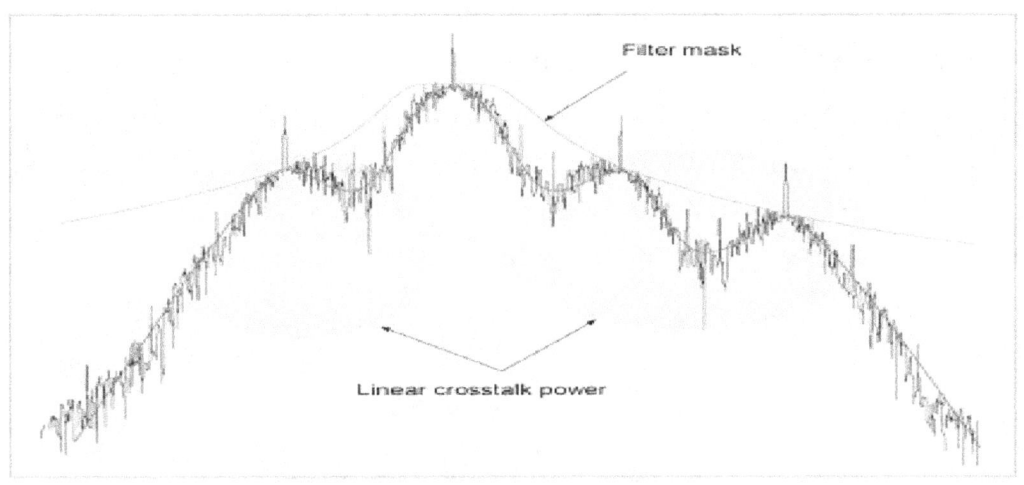

Filtered signal spectrum

Interference: Light can be imagined as a wave. The interesting question now is what happens if two of those waves meet? . The answer is interference. If the two waves have the same phase (that means „mountain to mountain and valley to valley ", the two waves add up and create a joint wave of higher amplitude. If the two waves have opposite phase though, they cancel each other, the result is „nothing ".

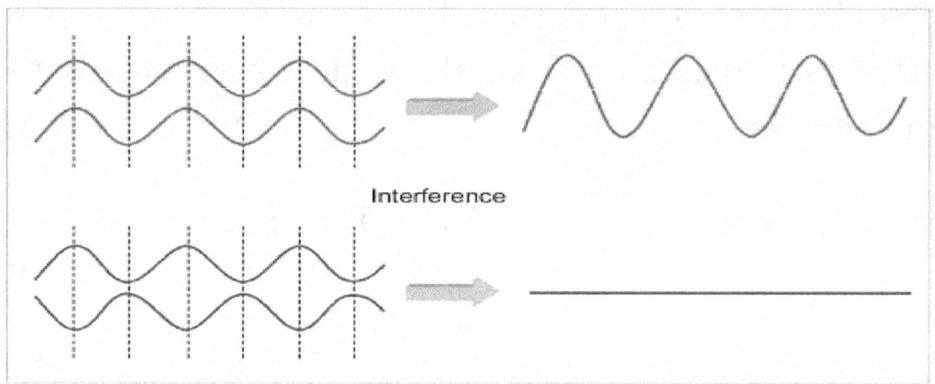

Interference

- **Fabry-Perot Interferometer:** The easiest form of interferometer is the Fabry-Perot type. It consists of two parallel plates that reflect light back and forth. By constructive and destructive interference only a few wavelengths are able to pass, the others are reflected. The criterion for constructive interference is that the differences in path length of the multiple-reflected beams is equal to an integer multiple of the wavelength. Thus, by varying the distance between the plates, certain wavelengths can be selected. For the other wavelengths the criterion is not fulfilled, therefore they are reflected.

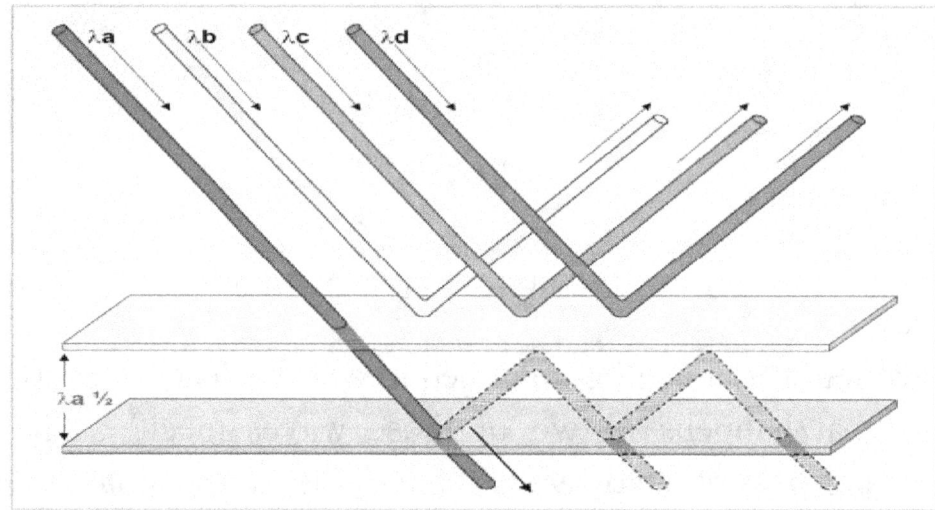

Fabry-Perot interferometer

- **Dielectric Thin Film Filters:** DTF Filters consist of alternate layers of high refractive index and low refractive index, each layer being $\lambda/4$ thick. Light reflected within layers of high refractive index does not shift its phase, while light reflected in layers of low refractive index is shifted by 180°. The condition for constructive interference once more causes one wavelength to pass and the others to be reflected. That means its function is similar to a Fabry Perot Filter but it is much more "accurate", with narrow line width etc.

Some features are:

- Low pass band loss: less than 0.3dB
- Good channel spacing: better than 0.8nm
- Low inter-channel crosstalk: better than –28dB

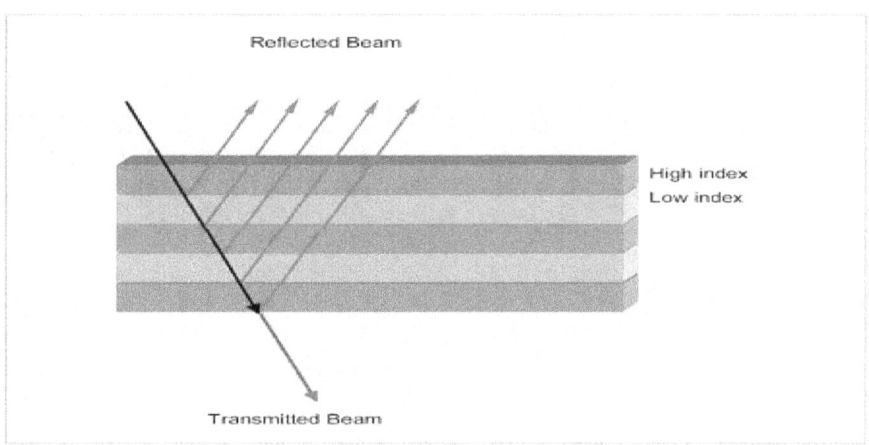

Dielectric thin film filter

- Bragg Grating: A Bragg grating (or Bragg Reflector) consists of a number of parallel semi-reflecting plates. Once more by using constructive and destructive interference just one specific wavelength is completely reflected, if it satisfies the condition

$$d = n*\lambda B/2$$

Where n is 1,3,5,...

Bragg reflectors have a very high reflectivity and are therefore employed as mirrors for high power lasers. A variation of Bragg gratings is the so-called fiber bragg grating: By varying the index of refraction of a fiber core it is possible to achieve a kind of Bragg grating, such that one wavelength is reflected, while the others pass through.

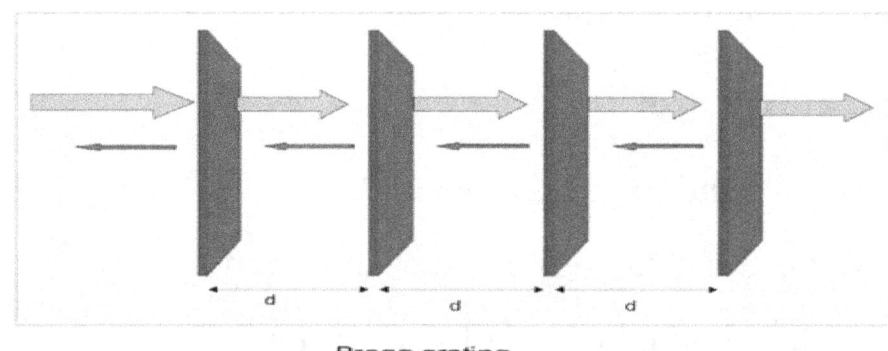

Bragg grating

Fiber Bragg grating

- **Mach-Zender Filter:** This method once more relies on interference. A mix of two wavelengths arrives at the first coupler which distributes the power equally on both lines. One of the lines is longer, thus introducing a different optical path length and a phase shift. Selecting that phase difference cleverly can mean that the first wavelength has its interference maximum at the place of fiber one and the second wavelength at fiber two, thus separating the two signals. By introducing a heating device to regulate the difference in length, and thus the phase shift, it is possible to tune a Mach-Zender filter.

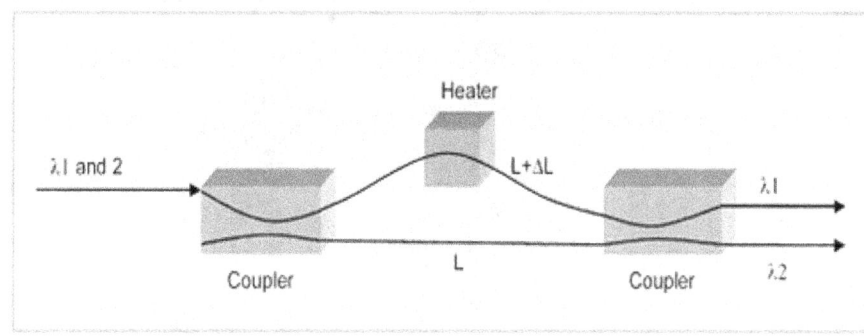

Mach Zender filter

3- Optical Multiplexer and Demultiplexer

An optical demultiplexer can be built as an association of optical filters or as a single-stand device. The purpose is to extract the original channels from a DWDM signal. The requested properties of this device are the same as for the optical filter: isolation and signal distortion. However, channel number and spacing must be considered now because demultiplexers can impose limitations on the number of channels or the total available bandwidth. Most demultiplexers are symmetrical devices and can also be used as multiplexers.

Components Prism: The easiest and best-known optical demultiplexer is the prism. Using the effect of dispersion (different speed of light for different wavelengths), light is split into its spectral components.

- Diffraction Grating: The function of a diffraction grating is very similar to that of a prism, only here interference is the important factor. A mixture of light is also split into its contributing wavelengths. With such a grating, sometimes also called a bulk grating, channel spacings of down to 50GHz can be achieved.

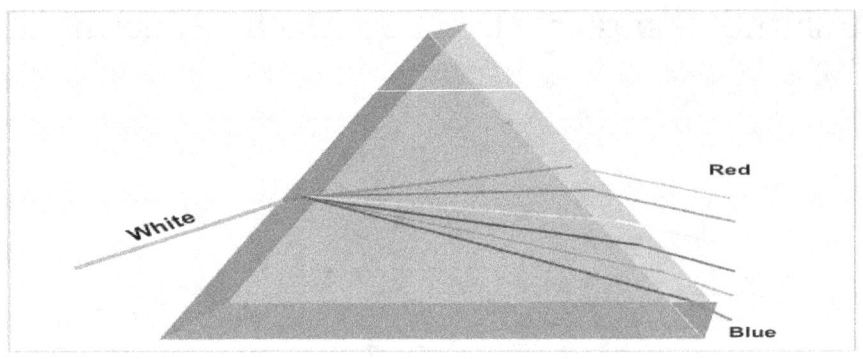

Effect of a prism

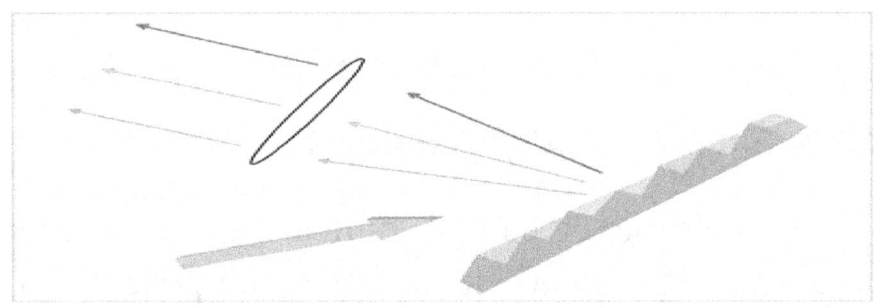

Effect of a grating

Dielectric Thin Film Filters: DTF filters are well suited to multiplex or de-multiplex a small number of channels. They are simply connected in sequence, each filter dropping one specific wavelength. Although this method is comparatively easy it has one drawback; each reflection causes attenuation of approximately 0.1dB. That means the channel de-multiplexed last is attenuated to a much higher degree then the first one, especially if we are talking about a higher number of channels. This property restricts the use of DTFs quite severely, limiting the number of channels to about 16. The minimum spacing reachable with these devices is about 100GHz.

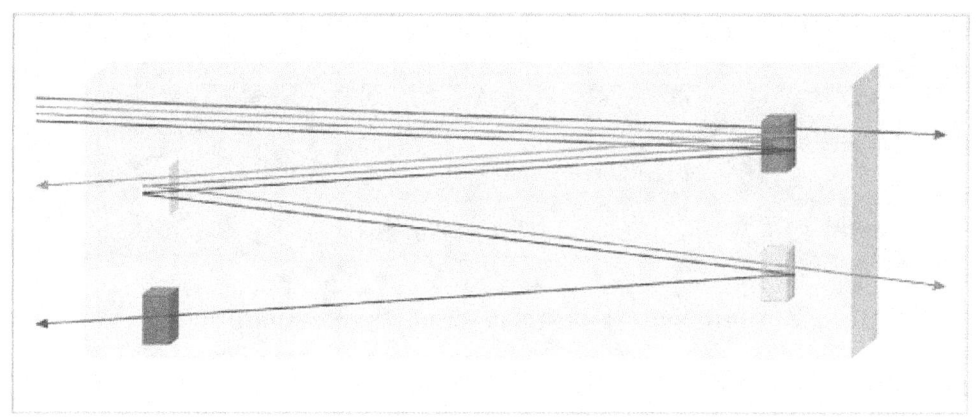

Demultiplexing using narrowband DTF filters

Arrayed Waveguide Gratings: Another technique also relying on interference are AWGs. Here, by introducing optical length differences it is ensured that each wavelength has a maximum at one of the output fibers.

The basic idea is very simple. For example, to execute the de-multiplexing of a WDM signal, the input signal is coupled into an array of planar waveguides after passing through a coupling section. During its propagation the signal in each waveguide experiences a different phase shift because of different lengths of waveguide. Moreover, the phase shifts are wavelength dependent which has once more to do with dispersion. As a result, different channels focus to different spatial spots or to the inputs of different waveguides. This method is better suited for a higher number of channels, as all channels suffer a more or less equal loss. The arrayed waveguide gratings can manage channel spacing of minimum 50GHz.

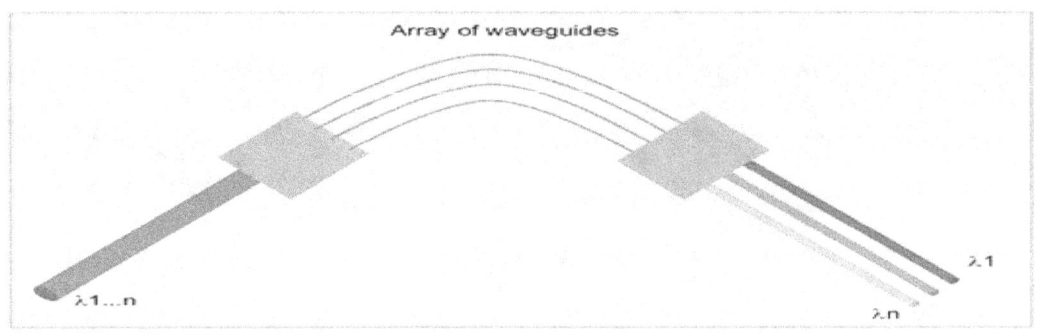
Functionality of an arrayed waveguide grating

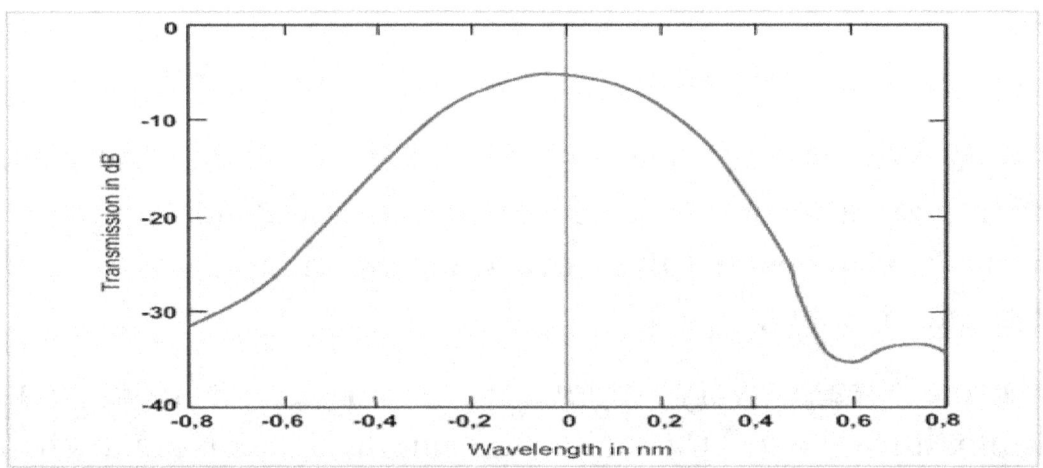
Example for the transmission of a channel of an AWG

Mach Zender Interferometer: Using a cascaded set of Mach Zender filters it is possible to demultiplex (or multiplex) a number of channels with different wavelength. Those Mach Zender interferometers can be integrated on Silica substrates, using conventional technology.

Fiber Bragg Gratings: To de-multiplex a multi-wavelength signal wavelength by wavelength it is possible to use a combination of an optical circulator and a Fiber-Bragg grating (FBG). This method is particularly interesting for optical add-drop multiplexers as single wavelengths can be easily dropped. Even more interesting is the possibility to tune that device by changing the length constant of the Fiber-Bragg grating using piezo technology. One advantage is the comparatively low insertion loss of only 0.2dB per FBG. A second one is the reachable channel spacing of only 25GHz.

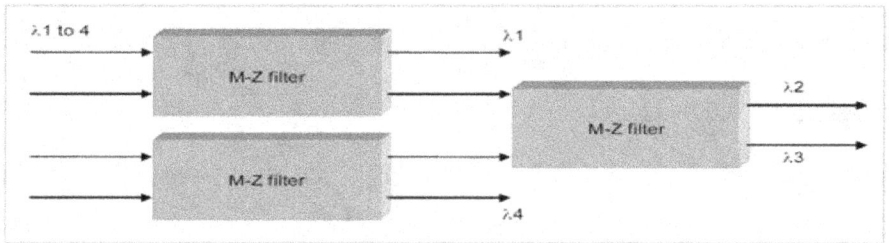

Functionality of a mach Zender interferometer

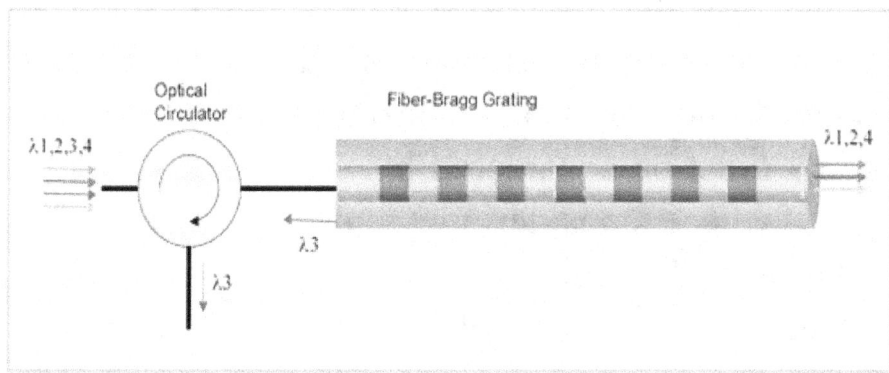

Fiber grating as demultiplexer

4- Optical Amplifiers

Fiber loss and dispersion limit the transmission distance of any fiber-optic communication system. For long-haul WDM systems this limitation is overcome by periodic regeneration of the optical signal at repeaters, where the optical signal is converted into electric domain by using a receiver and then regenerated by using a transmitter. Such regenerators become quite complex and expensive for multichannel light wave systems. Although regeneration of the optical signal is necessary for dispersion-limited systems, loss-limited systems benefit considerably if electronic repeaters were replaced by much simpler, and potentially less expensive, optical amplifiers which amplify the optical signal directly. Several kinds of optical amplifiers were studied and developed during the 1980s. The technology has matured enough that the use of optical amplifiers in fiber-optic communication systems has now become wide spread Optical Amplifier Applications:

- In-line amplifiers
- Booster amplifiers
- Pre-amplifiers

In-line amplifiers are used to directly replace optical regenerators. Booster amplifiers are used immediately after the transmitter or multiplexer to increase the output power. Pre-amplifiers are used before the receiver or de-multiplexer to increase the received power and extend distance. The use of each configuration as advantages and disadvantages that must be considered by the systems designer. The problems come when considering non-linear effects in the transmission fiber and also noise generated by the amplifiers. Some of the requirements for optical amplifiers for DWDM purposes are:

- high gain
- low noise
- flat amplification profile

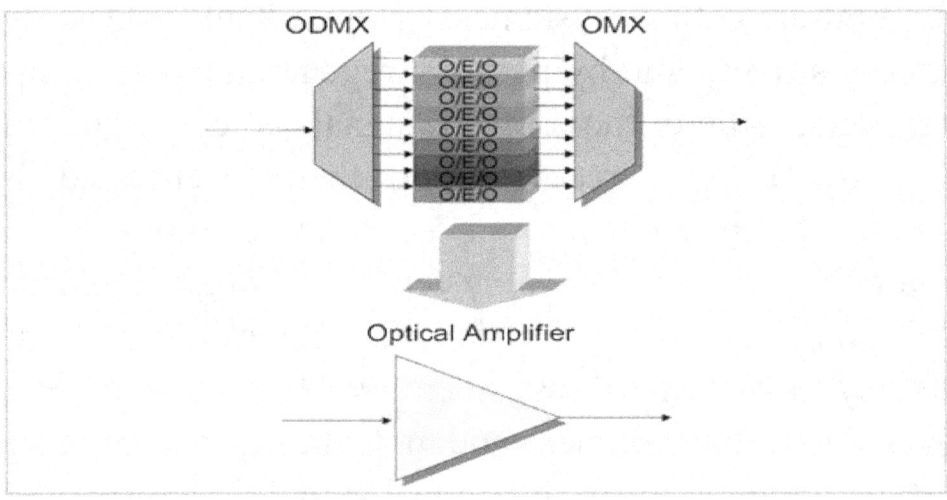

Passage from optical/electrical regenerators to optical amplifiers

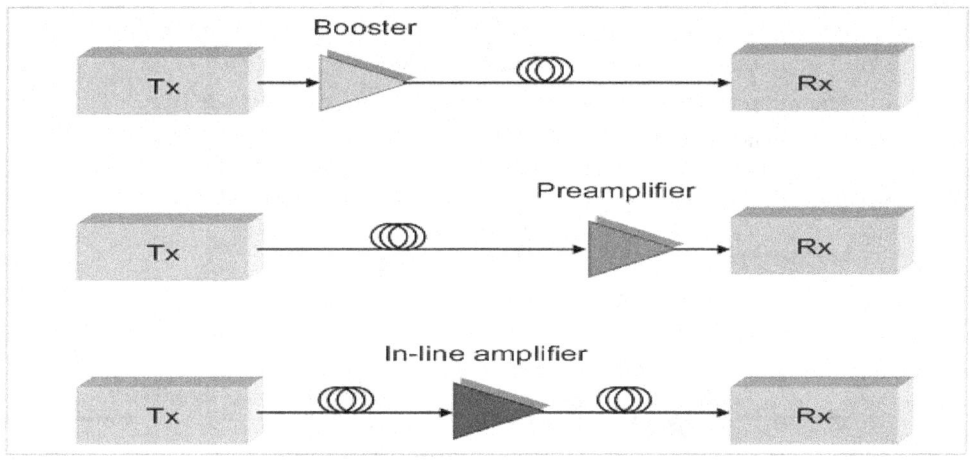

Applications for optical amplifiers

Erbium Doped Fiber Amplifier: For optical amplification purposes very often so called EDFAs (Erbium doped fiber amplifiers) are used, as well for pre-amplifying or boosting a signal. Those systems basically consist of a pump laser, a coupler device to join signal and pump light and an erbium-doped piece of fiber of several meters of length. A pump laser at between 980 and 1480nm lifts electrons of the Erbium ions from the energy E1 to a higher, excited energy level. In the next step the electrons fall down a little to the energy level E2. In that way itis ensured that pump light and signal light don't influence each other. Finally, the phenomenon of „induced emission "is used. A photon of our signal light passes the electron; consequently, that electron falls down to its former energy E1, at the same time emitting another photon identical to the one which has just passed. Thus, we have amplified our signal. To put it briefly, in the EDFA the pump light is converted to signal light, in that way amplifying the signal. The gain is influenced by factors like erbium-ion concentration, core radius, amplifier length, pump power and pump configuration. The EDFAs can be used to amplify the whole third and fourth transmission windows typical values for EDFAs are:

	C-Band	L-Band
Wavelength range	1530-1565nm	1570-1605nm
Total output power	14-25dBm	14-25dBm
Length of active fiber	10-60m	50-300m
Number of pump lasers	2-6	3-8

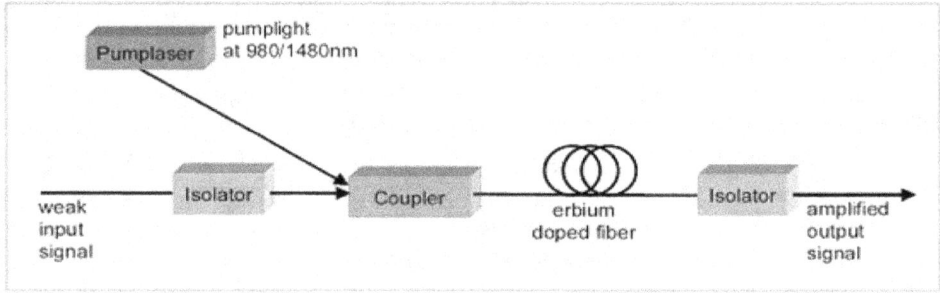

Layout of an EDFA

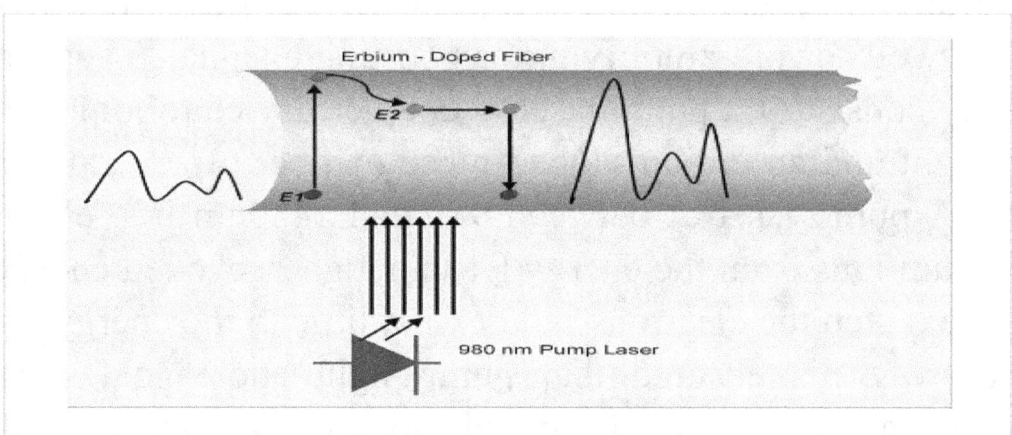

Functionality of an EDFA

Disadvantages:

- The amplification is not linear over the range. In order to adjust all channels to a comparable power level (compensate so called tilt) it might be necessary to use variable optical attenuators (VOAs) to pre-emphasize each individual channel.

- A second problem is that (in the easiest implementation) the total output power is kept constant, independent of the number of channels transmitted. That means, the more channels, the less power per channel and the shorter the possible hop

- Problem number three is the so called amplified spontaneous emission (ASE). Some of the excited electrons fall down to ground level without being induced. That radiation is also amplified and causes a good part of the noise background.

Advantages:

- Simultaneous amplification of the whole 1550nm area
- Total output power 1 to 1000mW possible
- Large dynamic ranges
- Suitable for long-haul applications
- Therefore, at the moment the EDFAs are the commonly used amplifier type.

Hints: EDFAs are often cascaded, giving two advantages:

- Noise reduction by e.g. cascading one amplifier pumped at 980nm and another at 1480nm.
- Bye adjusting the characteristics of the amplifiers with VOAs the tilt can be compensated.

Like Erbium other rare-earth elements can be used for amplification in other wavelength ranges. At the moment those amplifiers are not ready for use, though.

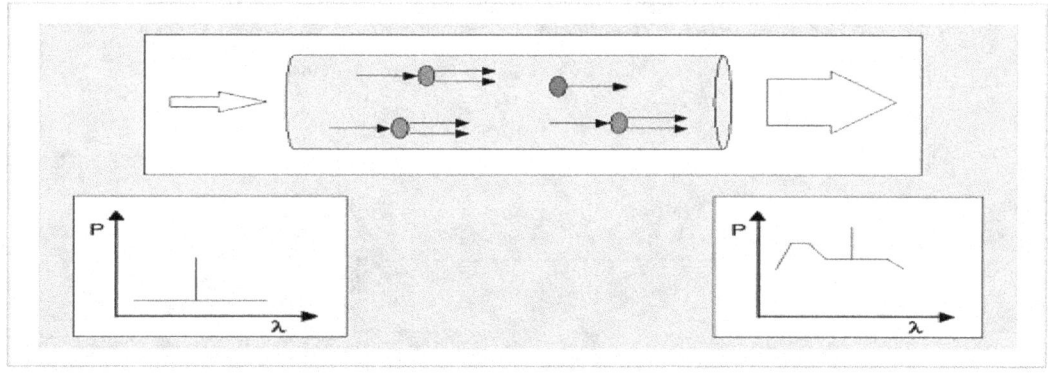

Generation of noise (ASE) in an EDFA

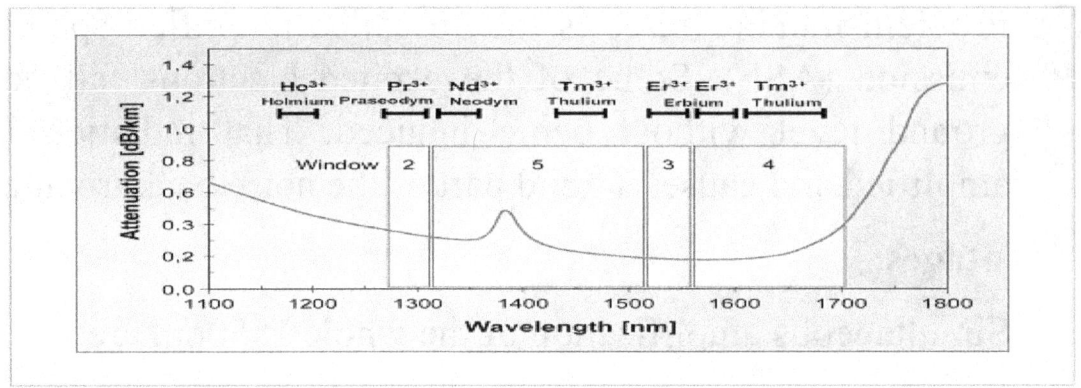

Materials used for fiber amplifiers

Raman / Brillouin Amplifier:

The Stimulated Raman and Brillouin Scattering can also be used for amplification. In contrast to the fiber amplifiers these amplifiers are often pumped from the receive side. One major advantage is that these amps can be used over the entire range from 1300 to more than 1600nm, depending on the wavelength of the pump light, which is by scattering transformed to signal light. The achieved gain can be in the range of 10 to 14dB.

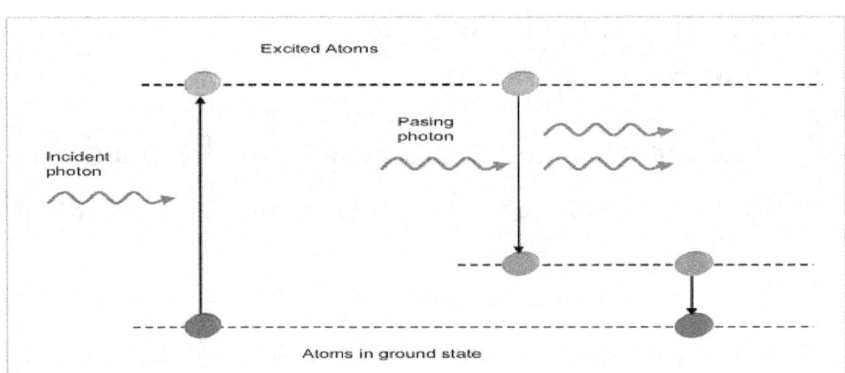

Functionality of a Raman amplifier

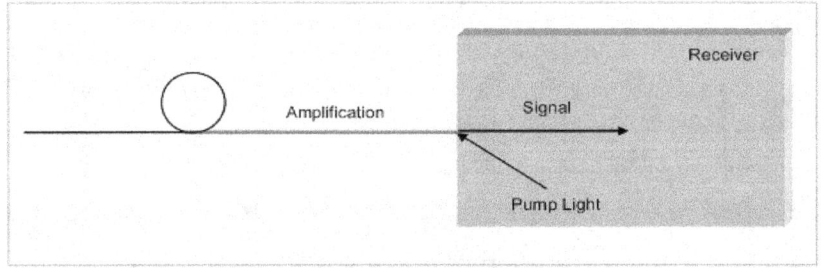

Layout of a contradirectionally pumped Raman amplifier

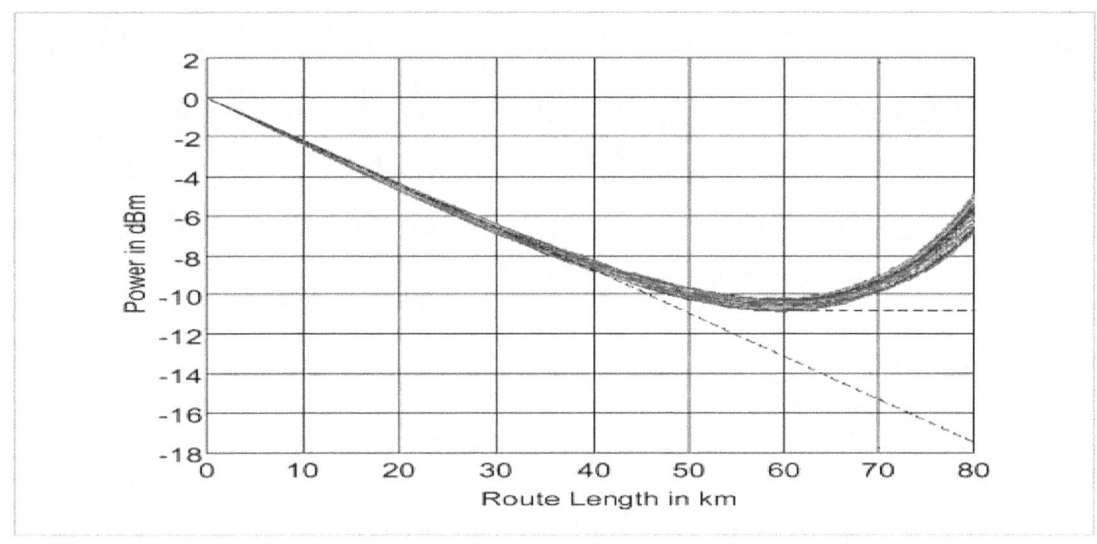

Typical power graph along a contradirectionally Raman pumped line

Wavelength stability is the laser's ability to maintain the same transmitted wavelength with environmental changes (i.e. temperature, humidity). Chirp or carrier frequency shift, are oscillations in the transmitted wavelength. These oscillations are due to the dependence of the laser material on the applied current. This effect occurs especially in directly modulated lasers and a practical result is the increase of the signal bandwidth.

It should also be noted, that lasers also serve as pump-lasers for EDFAs. Those lasers are of a rather high power and often are cascaded via different coupler devices to give the required high pump power.

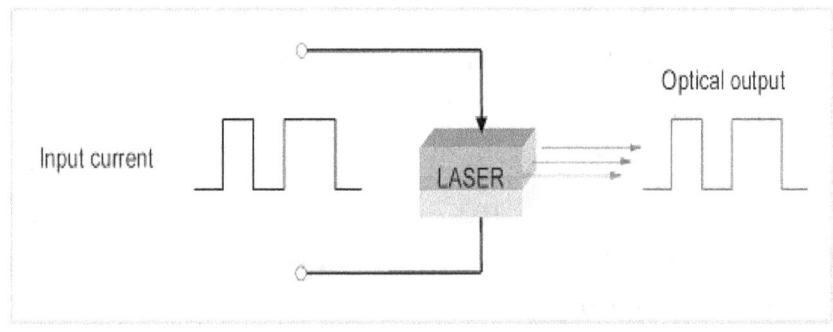

Laser configuration with direct modulation

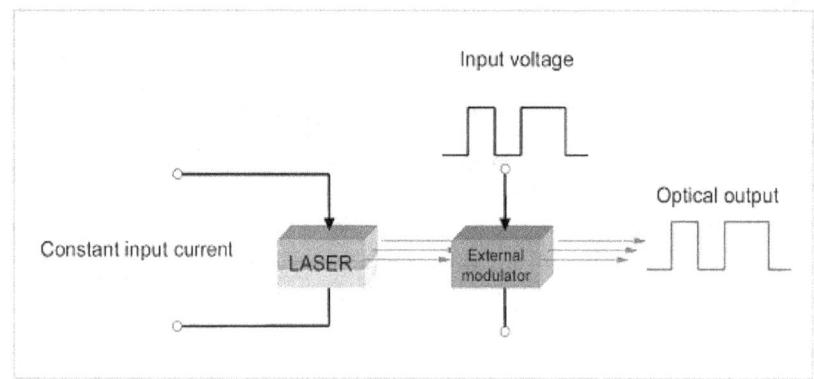

Laser configuration with external modulation

- **Fabry-Perot-Laser:** The easiest form of a semiconductor laser is the Fabry-Perot-Structure:

Basically, these lasers are planar light emitting diodes consisting of a n-type and a p-type doped layer. Between these two layers, we've got the active region where the takes place. To be able to select a lasing wavelength this region has to have the form of an interferometer, which in this case can be done by cleaving. The length of the lasing cavity determines the lasing wavelength. Like a Fabry-Perot Interferometer, the FP Laser is not very "good": In addition to our desired wavelength we also get "side-wavelengths" which effectively give us a larger linewidth. Therefore, the FP Laser is also called "Multi-Longitudinal-Mode" (MLM) Laser.

- **Distributed Feedback Lasers:** DFB Lasers are commonly used for DWDM purposes. Here, we're applying basically the same trick as in the Dielectric Thin Film filter: We don't use simply mirrors at the end, but introduce a layer-structure in the lasing cavity, a bit like in a Bragg Grating. This has a similar effect like a FP structure, just much better. While having a slightly more complicated structure, DFB Lasers fit the requirements very well:
- Narrow peaks (about 0.0001nm)

- Wavelength range 1520 to 1565nm and above (third and fourth window)
- Stable

As they really only produce one lasing mode, the DFB lasers are also known as SLM (Single Longitudinal Mode) Lasers.

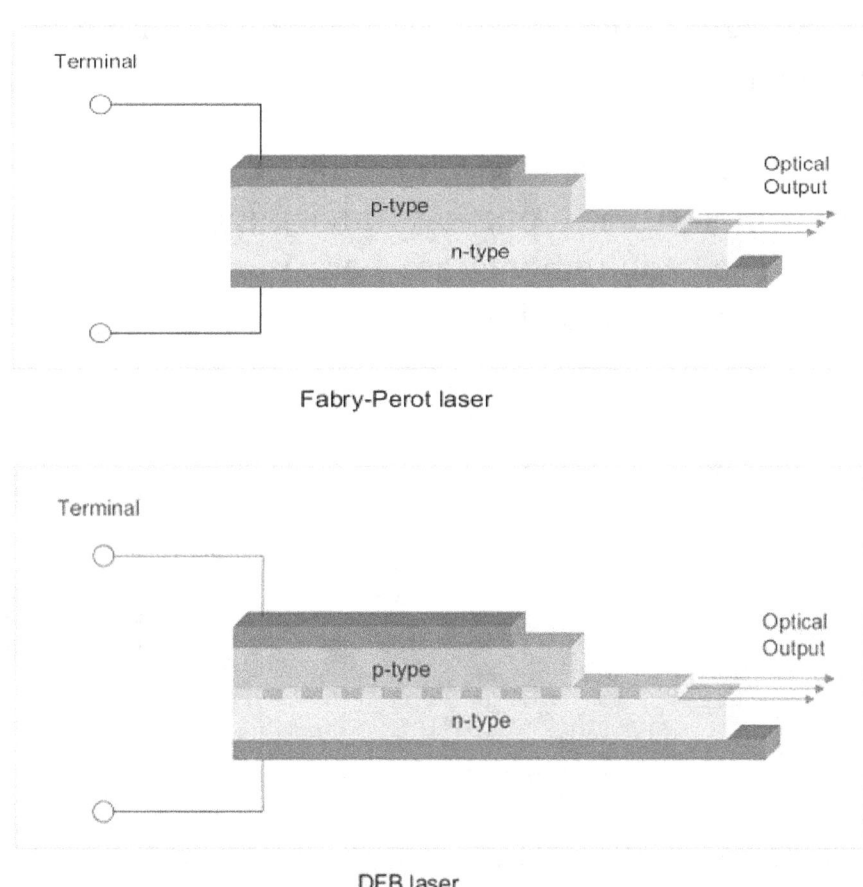

Fabry-Perot laser

DFB laser

- **Comparison**: While FP Lasers are quite commonly used in usual single channel transmission as sources of "grey" light, DFB lasers are the common laser source for DWDM systems, due to there much smaller linewidth.
- **Tunable Lasers:** The newest development in laser technology is tunable semiconductor lasers. Those lasers are highly desirable as they have a lot of features which are interesting for DWDM:

  ➢ One laser for several colors (spares!)

> use in OADMs

> use in optical cross-connects

> use for optical protection switching, etc.

From the technical point of view lasers can be tuned by varying the refractive index of the lasing cavity. Changing that refractive index is equal to changing the length of the cavity and thus the selected sending wavelength.

- **Pump Lasers:** For the pumping of EDFAs pump lasers are used, which usually put out rather high powers, up to the watt range. The requirements are here more polarization properties and power then linewidth

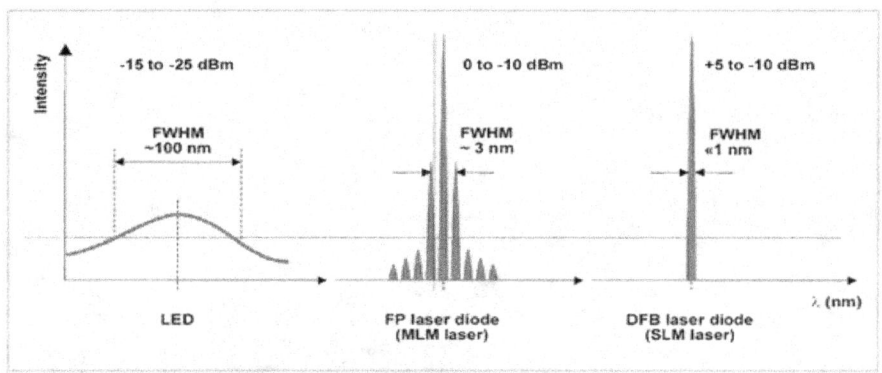

Properties of different LED and laser types

- **Modulators:** The modulation of lasers can be done by modulating the laser itself or using an external modulator. Direct modulation brings the problem of „chirping ", i.e. frequency oscillations of the laser. The reason for this is that the index of refraction of the active region changes slightly with the applied current, therefore the effective length of that region changes and also the emitted wavelength. As this is a problem in DWDM systems external modulators are often used. They solve the problem of chirping, butthey cannot avoid a certain broadening of the line, which is physically inevitable when modulating a signal.

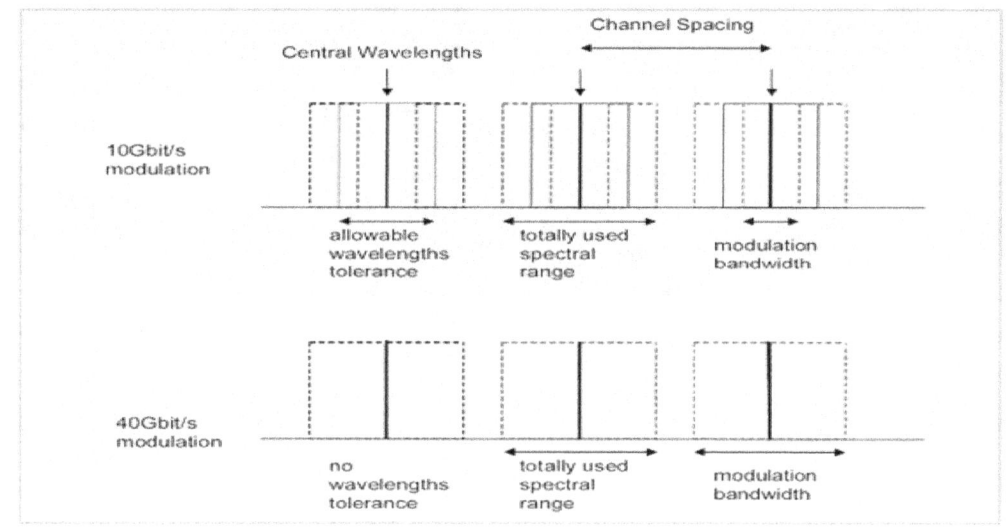

Effect of modulation on the channel bandwidth

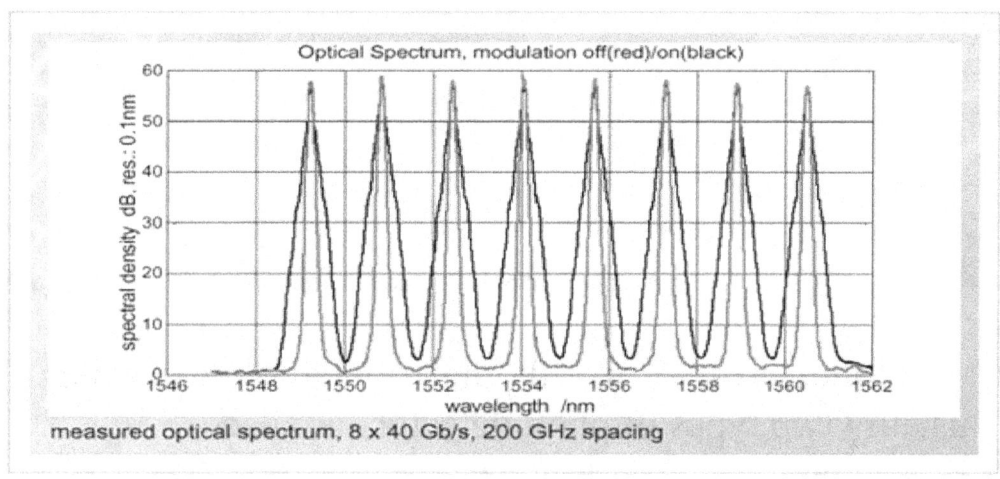

Spectral broadening by modulation

- **MQW Modulator:** One type of modulator is the semiconductor MQW (Multiple Quantum Well) modulator. Its functionality is simple. when voltage is applied, light is absorbed. It has the further advantage that it can be produced on the same substrate as the laser itself.
- **Mach Zender Modulator:** Its function is analogous to that of the Mach Zender filter. By varying the phase of one arm of a M-Z filter the two parts of the signal either interfere constructively ("on") or destructively ("off"). $LiNbO_3$ can be used for this phase control.

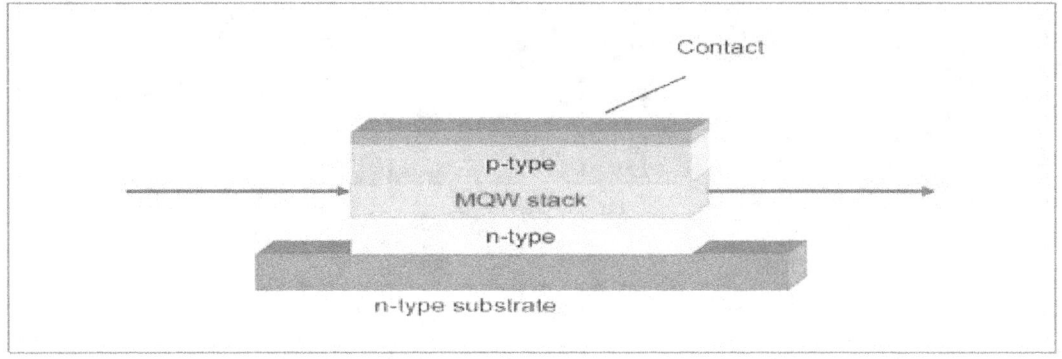

Semiconductor MQW modulator

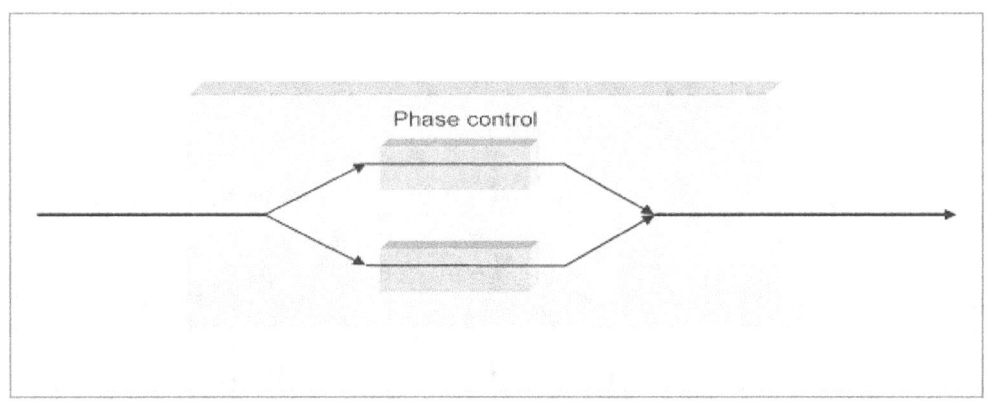

Layout of a Mach Zender modulator

- **Photodetectors:** In all the optical transmission systems, usually two types of photodetectors are used:
- PIN-Diodes
- APD Diodes

  > **PIN Diode:** The name comes from the structure of the diode: p-doped, intrinsic and n-doped Semi-conductor material is used in a layer structure. The diodes are reversely biased. Each incident photon causes an electron-hole pair to be produced, the electron and hole drift towards the electrodes which in turn a measurable current. This current is proportional to the number of incident photons. The intrinsic layer has been inserted, in contrast to usual p-n diodes, mainly to reduce distortion.

- **APD Diode:** In principle the APD (Avalanche Photo Diode) works the same way as the PIN diode. The difference is that the electrons and holes, while being separated, gather enough energy to cause more electron-hole pairs by impact ionization. Thus, a whole "avalanche" of charged particles flows and a higher current can be measured. Therefore, APDs are especially well suited for applications where a very high sensitivity is needed. It should be mentioned though that the avalanche process is rather "noisy", causing a fluctuation of the gain factor.

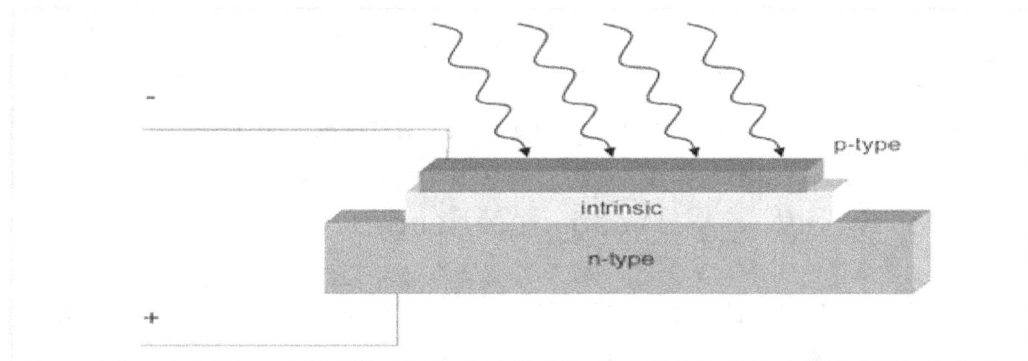

Layout of a photodiode

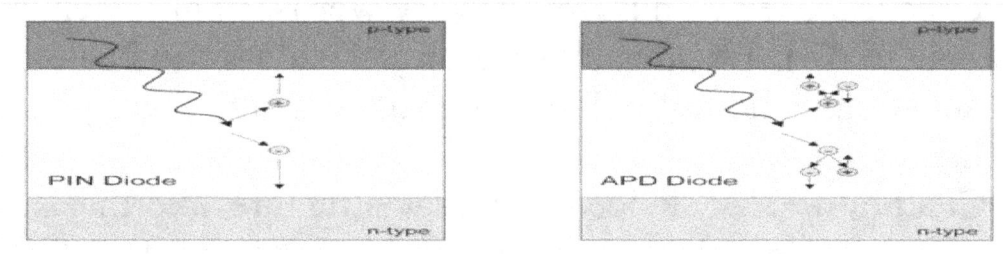

Functionality of PIN and APD diode

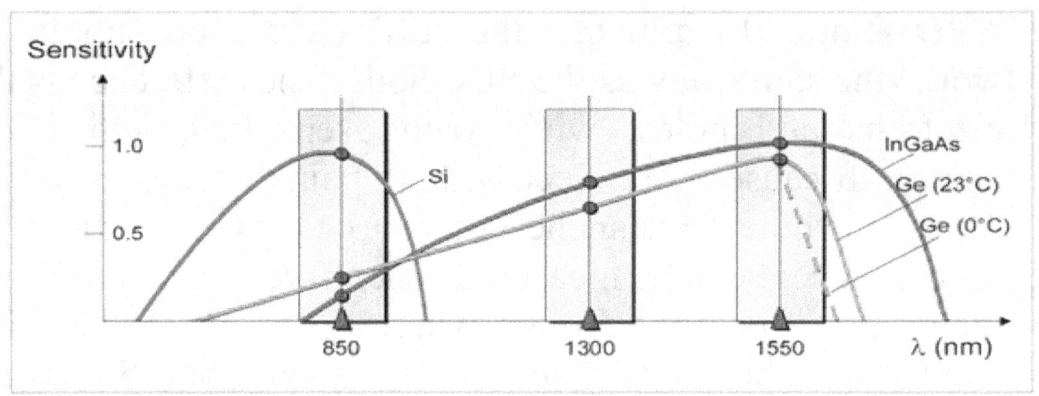

Spectral response of different detector materials

- **Isolators, Circulators and Connectors:** The very-narrowband lasers used in DWDM are highly sensitive to reflected power. Such external reflections act as additional resonators and the result might be wavelength instabilities, mode jumps, noise, etc. Even a power 6 orders of magnitude lower than the transmitted power (about -60dB) can significantly disturb the laser.

  ➤ **Isolators:** Isolators are, to put it simply, devices that let light pass in one direction without attenuation and do not allow light to flow in the reverse direction. In that respect they are a kind of "optical diodes". Thus, reflected power is highly attenuated (by about 30dB). Technically speaking, isolators can be constructed as a combination of polarization rotators and linear polarization filters.

  ➤ **Circulators:** A device of similar structure like the isolator is the circulator. It works as a kind of multiport isolator, transmitting the input of port 1 to port 2, input of port 2 to port three and so on.

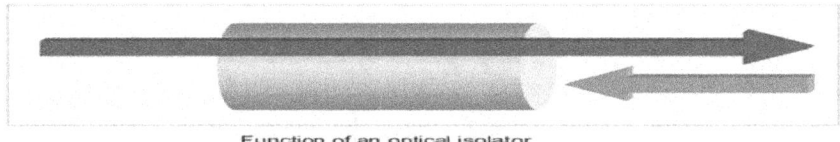

Function of an optical isolator

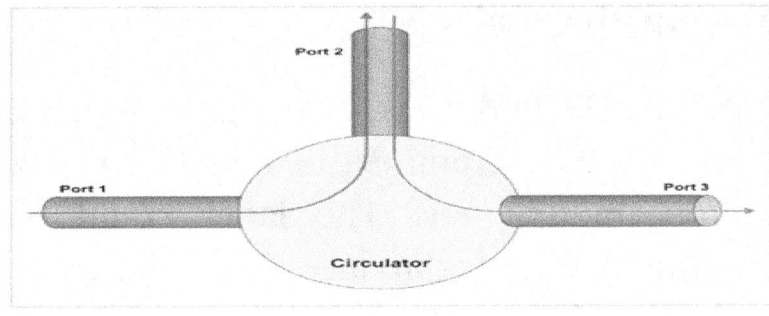

Optical circulator

- **Connectors:** Connectors have to be especially selected to guarantee low reflection. They should satisfy two criteria:

    ✓ Physical contact
    ✓ Angled contact

Types like E2000 or /APC or, generally speaking, high-return-loss connectors are therefore good choices. Additionally, all connectors have to be cleaned very carefully to avoid breaks in the physical contact and also to avoid power being absorbed in those impurities.

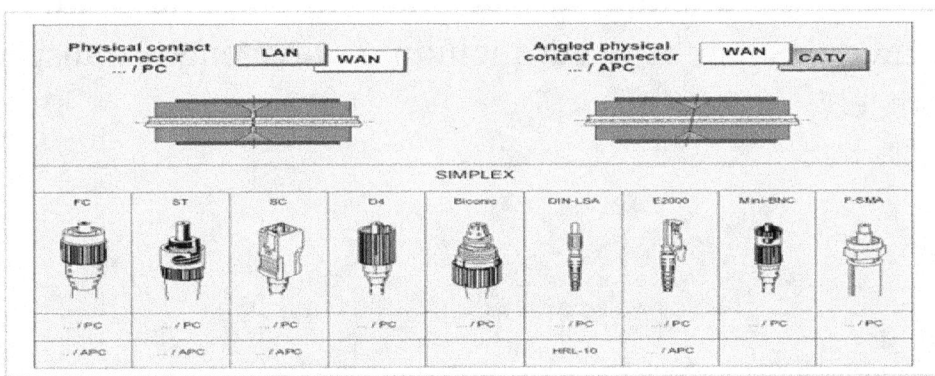

Connector types

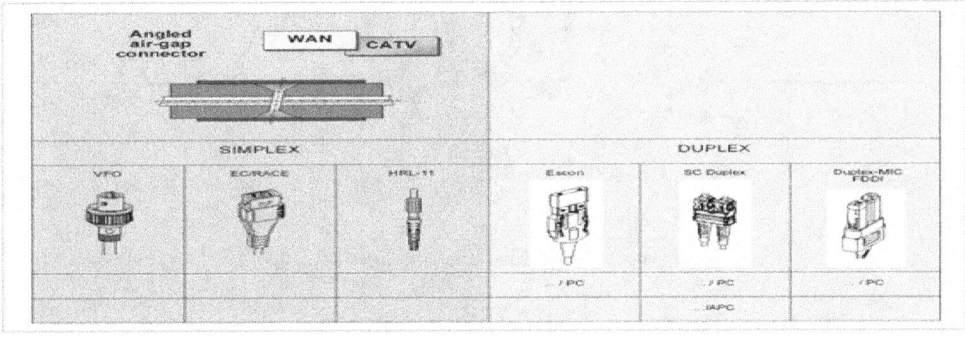

Connector types

5-Optical Switching Units

Not really a part of DWDM lines, but of growing interest for the all-optical network is the development of optical cross connectssand of optical add-drop multiplexers. The latter could be implemented already with a number of techniques discussed above, like tunable fiber-bragg-gratings, DTFs or tunable lasers. Newer developments include more sophisticated and cheaper solutions. The really big optical cross-connect remains the challenge though. At the moment optical traffic is connected with the help of distribution panels, within the next years is will be necessary to switch at least part of the exploding data traffic via quickly reconfigurable optical cross connects. Basically, it is possible to switch the traffic either electrically or optically. The electrical switching fabric is only capable of handling comparatively low bitrate traffic; therefore, we will concentrate on the optical variants. Generally speaking, larger switching fabrics can be created by cascading smaller units in one way or another. In addition to the switching fabric a number of additional modules can be used, including wavelength transponders, regenerators, etc.

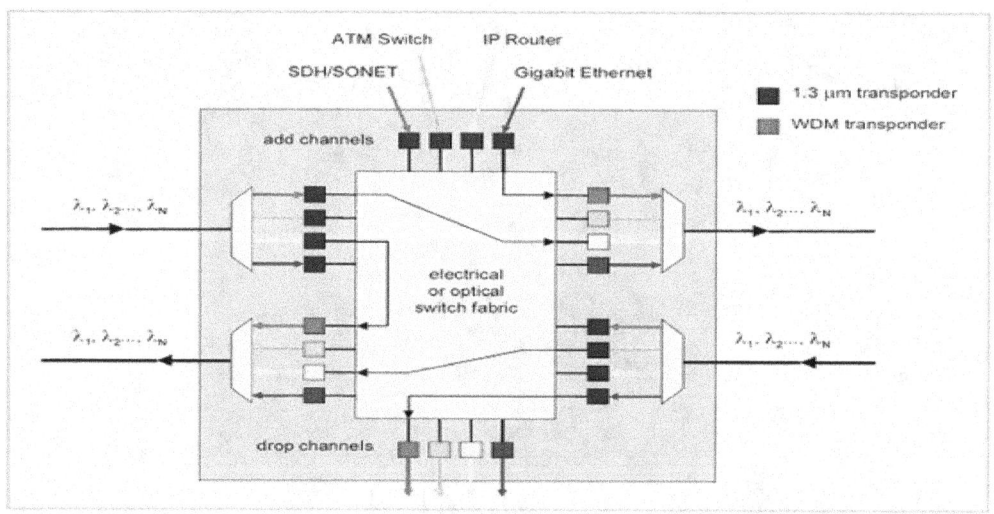

Functionality of an optical crossconnect

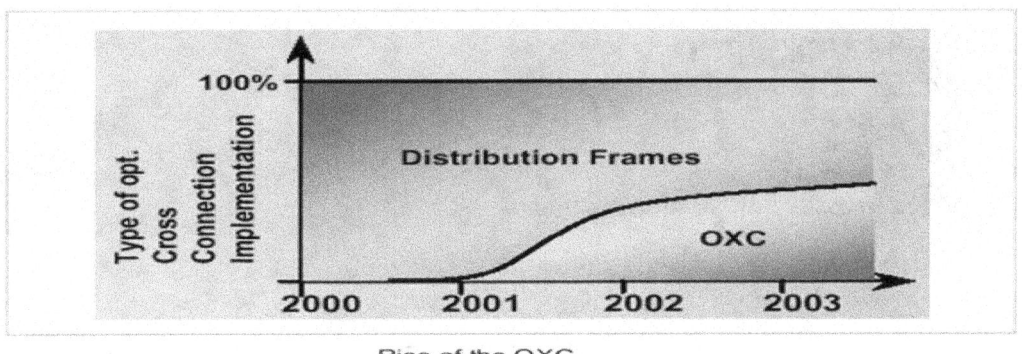

Rise of the OXC

At the moment quite a few different technologies are used on the way to the all optical switch or the optical add-drop multiplexer. Differences are e.g. the maximum size of the matrix or the switching time. Among those methods are:

- **Solid state cross-connects:** By changing the physical properties of semiconductor directional couplers two incoming signals are switched to two outgoing fibers. By cascading those 2x2 switches larger switching matrixes can be generated. Switching speeds in the nanosecond range can be achieved.
- **Liquid crystals:** Borrowed from lap-top screen technology, electric currents alter the properties of liquid crystals so that light passing through them is polarized in different ways. Passive optical devices then steer each wavelength of light one way or the other, depending on its polarization.
- **Tiny bubbles:** These act like mirrors, glancing light onto intersecting paths as they traverse microscopic troughs carved in silica. The bubbles are generated using ink-jet printer technology.
- **Thermo-optical switches:** Light is passed through glass that is heated up or cooled down with electrical coils. The heat alters the refractive index of the glass, bending the light so that it enters one fiber or another. The same can be done using polymer technology.

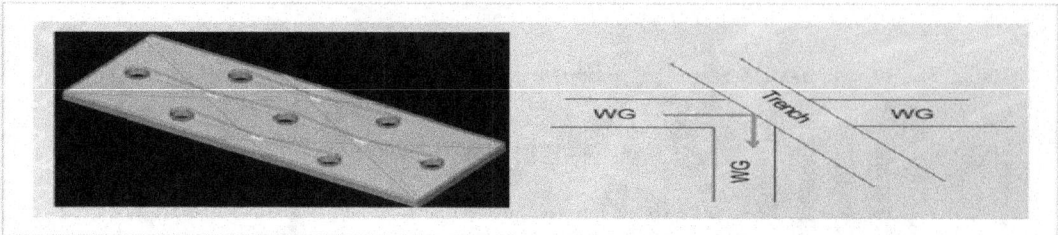

Use of bubbles for switching

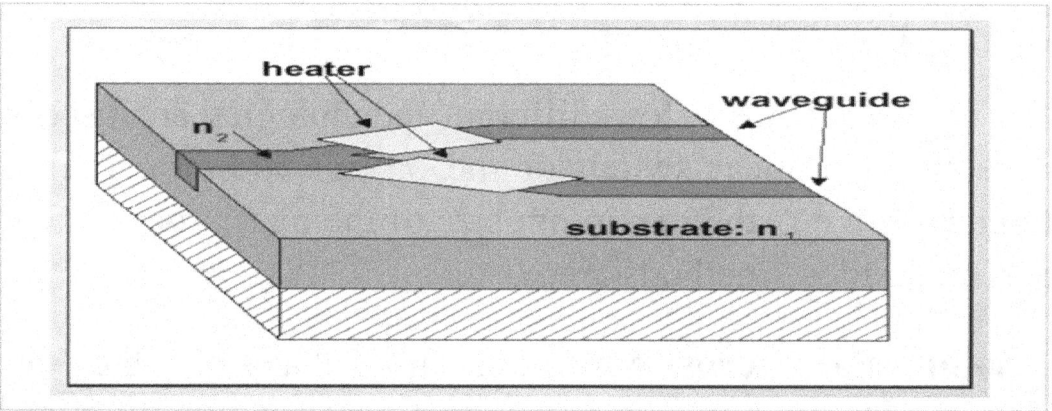

Thermo optical switches

- **Tunable lasers:** These devices pump out light at different wavelengths, and can switch from one wavelength to the other very quickly. Thus, they are a good solution for use in many different DWDM NEs, including switches, OADMs but also usual terminal optical multiplexers.

- **MEMS (micro-electro-mechanical systems):** Arrays of tiny mirrors originally developed for the very large video screens seen at sports events and pop concerts. They are used to reflect the light of one fiber to the selected output fiber. These mirrors orprisms can be integrated on a chip.

Advantages are e.g. a large possible scale of the matrix (1000x1000 is under discussion), low loss connectivity and compact design. Switching speeds are in the order of microseconds or milliseconds. Some facts that have to be taken care of are mechanical stability and long-term reliability.

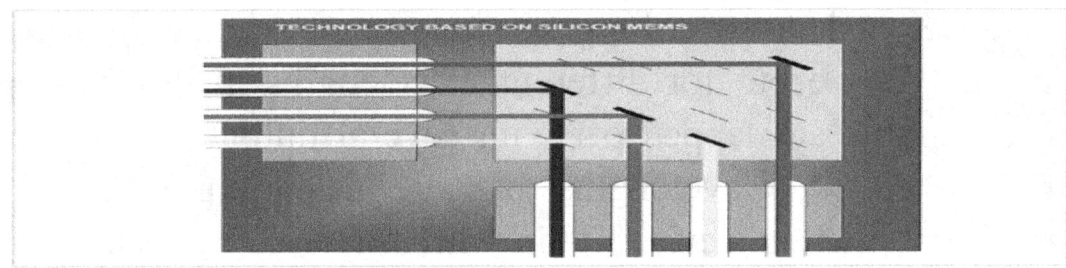

Functionality of a MEMS module

6- OSC (Supervision and control)

- SDH use number of bytes in overhead called (DCC) channels for supervision.

- WDM used one optical channel (wavelength) transfer with the other wavelengths and separated from other channels (traffic) at the receiver and it usually 1510 nm.

DWDM Network Element Structure:

In the past years a number of breakthrough new optical technologies made its way into commercially available components for DWDM networking equipment. "ROADM" is a buzzword, used by everybody, with better or worse understanding of the gives and takes, of the advantages and disadvantages.

Today two types of fixed filter technologies are used:

- Arrayed Waveguide Grating (AWG), and
- Thin Film Filter (TFF).

Both technologies are mature and on the market since several years. AWG based fixed filter are the better choice for most applications, TFF are still used for low port count metro access applications.

1. AWG filter technology:

AWG filter technology allows putting a high number of channels on one single silicium device, see Figure. The incoming signal is distributed in the lens area to all delay lines. Since the delay lines have different lengths, all signals interfere in the mixing area, and signal maxima occur for different wavelengths in this mixing area. These maxima are tapped off with several output fibers. With this setup up to 40 wavelengths can be de-multiplex in a single monolithic silicium device. Thin film filter (TFF) technology in a thin film filter, on the contrary, one single stage reflects only one single wavelength, which can be coupled out of the WDM signal. To design filters with several output ports, several thin film filter stages have to be cascaded. Typical thin film filter-based devices have 4 to 20 ports.

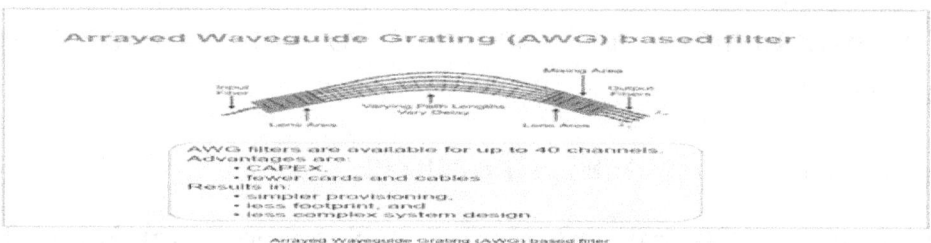

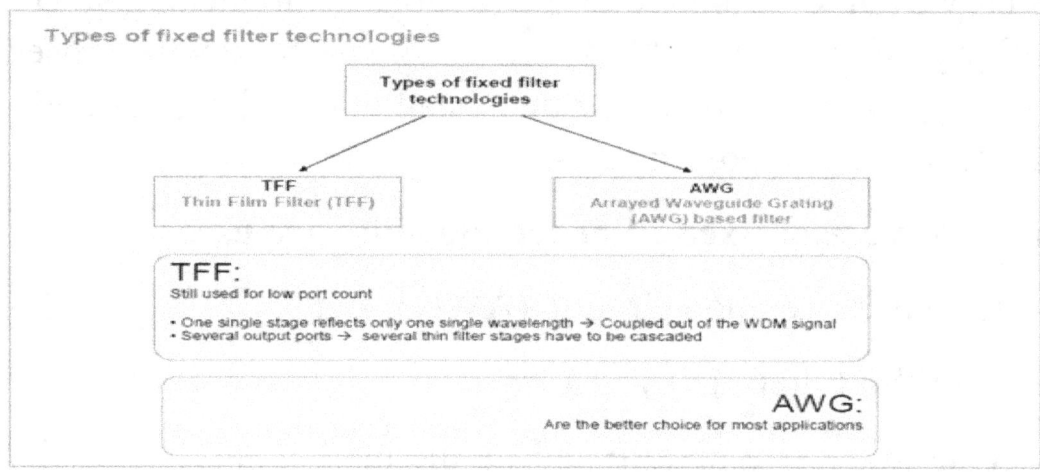

Types of fixed filter technologies

Serial fixed filter OADM:

A serial filter OADM – called "small OADM" is a good low-cost choice for low drop channel counts, where traffic patterns can be well predicted. In the serial setup multiple fixed filters can be cascaded in series to drop more than just one wavelength. The total number of cascaded filters is limited by the insertion loss and system design.

Such serial fixed filter architecture is best suited for small OADMs, where only a limited number of channels has to be dropped, typically up to ~8. The architecture is advantageous from a CAPEX point of view, because the necessary filter set-up has low cost. The low insertion loss limits the amount of necessary amplification, which again reduced the cost for EDFAs. It is often possible to use only preamplifiers (amplifiers on the receive side of the OADM), and to work without booster amplifiers the amplifier on the transmit side).

Disadvantages: The flexibility is limited: once a serial fixed filter OADM is installed, the choice of added/dropped wavelength cannot be changed without interrupting the express traffic, i.e. the wavelength channels expressing through the OADM. The banded architecture can "burn" wavelengths: if a band of four wavelengths is dropped, but actually only one wavelength is needed, the other 3 wavelengths cannot be used anymore

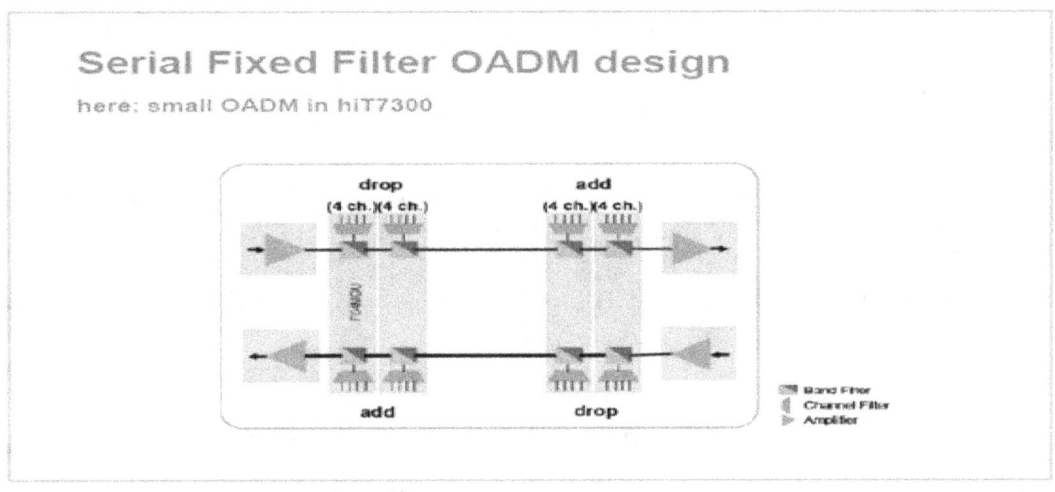

Serial Fixed Filter OADM design

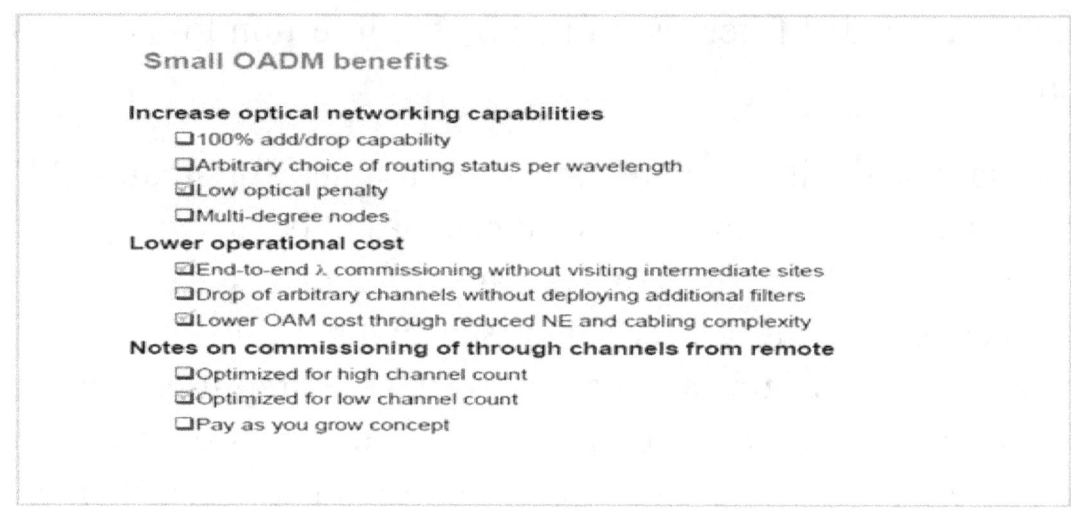

benefits

Parallel fixed filter OADM: The parallel fixed filter OADM approach follows a different philosophy: it de-multiplexes the DWDM signal down to all individual bands or even individual channels. The architecture is actually the same as for a terminal, only that two mirroring filter structures are used to de-multiplex and multiplex the channels.

Parallel fixed filters can either use:

- many individual 4 channel filters, "flexible OADM", or
- large AWG filters, "full access OADM".

The advantage of the first one is its flexibility, specifically for meshed networks. The advantage of the second one is its simplicity two different filter technologies can be used, each of these has its own merits.

Flexible OADM: First, band filters in combination with small – typical 4 channel – channel filters can be used to break down the complete DWDM signal in several steps down to the individual channel level. See the figure below for an example. Since this stepwise approach allows for more flexibility than the second approach, we call this architecture the "Flexible OADM".

Full access OADM: Secondly, so-called AWG (Arrayed waveguide Grating) filter technology can be used to break down a complete 40 channel signal into the individual channels in one step. See the second figure below for an example. Since the architecture gives access to all individual channels in just one filter step, we call this architecture the "Full Access OADM".

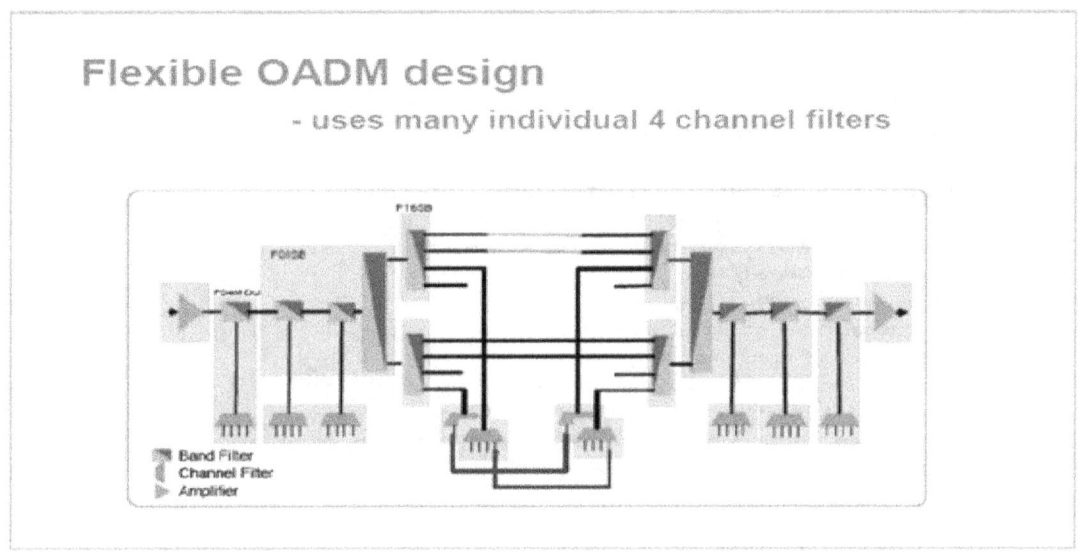

Flexible OADM design

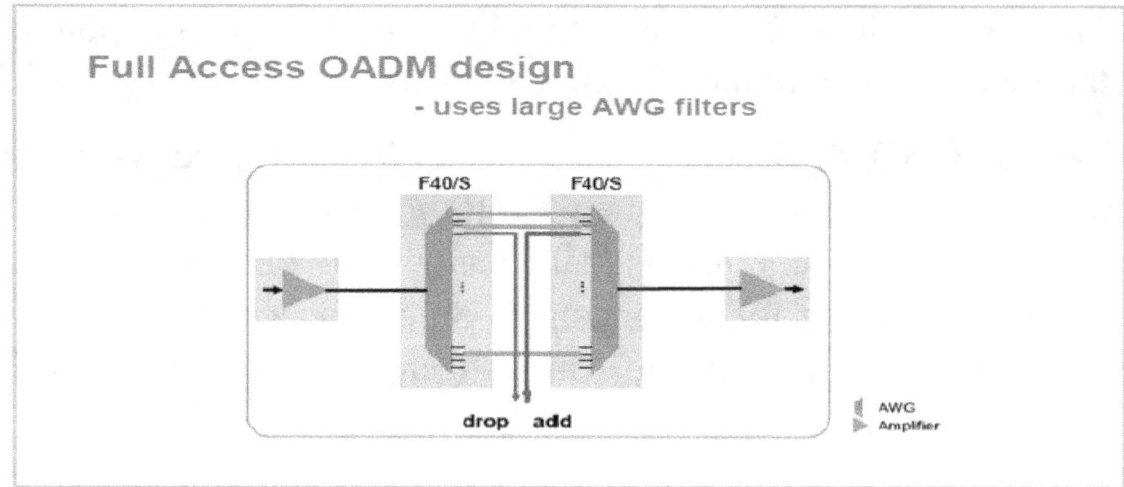

Full Access OADM design

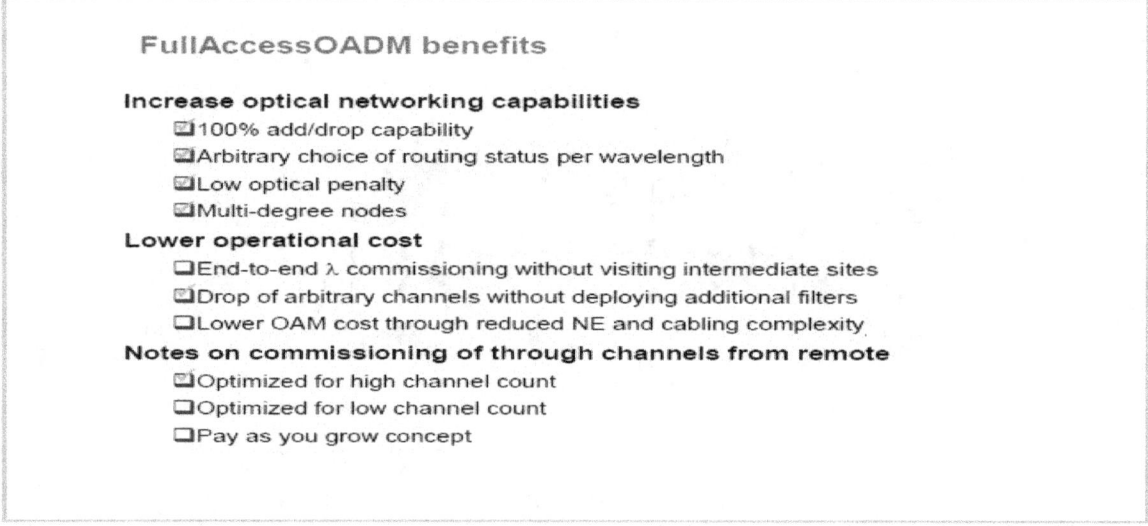

benefits

ROADM benefits: There are various benefits a carrier can expect from installing ROADMs in his WDM Network:

The main ROADM benefits are:

- simpler node installation,
- faster wavelength provisioning,
- simpler trouble shooting,
- increased reach, and

- resilience on wavelength level

Simpler installation: A ROADM or PXC node needs much less cables than a fixed OADM node, because many of the cables can be substituted by the ROADM card. all the (up to 40) patch cards per direction, which connect the two input and output AWG filter can be replaced by just one fiber pair connecting the add and the drop cards. This means also that all ROADM nodes look alike, have same cabling and card setup, regardless of the traffic matrix. A fixed OADM node looks different in terms of cables, cards, card placement for different traffic matrices.

Remote wavelength provisioning: The optical switch element in the ROADM allows switching wavelengths remotely. Switch actions can be to include an optical by-pass from remote, or to change in general the switching state. Switching into by-pass from remote is particularly attractive, because this allows to avoid truck-rolls altogether for all intermediate nodes end route two nodes, between which a new service is provisioned. Therefore, the process of switching becomes much easier with ROADMs, compared to fixed OADMs. This allows carriers to guarantee much faster wavelength provisioning times to their customers, which can turn into a valuable competitive edge in the competition for customer business.

Lecture 5
Access Networks

5.1 ADSL Modems

Applications of broadband Access:

Residential Application	Downstream rate (kb/s)	Upstream rate (kb/s)	Willing to pay	Demand Potential
Database Access	384	9	High	Medium
On-line directory; yellow pages	384	9	Low	High
Video Phone	1,500	1,500	High	Medium
Home Shopping	1,500	64	Low	Medium
Video Games	1,500	1,500	Medium	Medium
Internet	3,000	384	High	Medium
Broadcast Video	6,000	0	Low	High
High definition TV	24,000	0	High	Medium

Business Application	Downstream rate (kb/s)	Upstream rate (kb/s)	Willing to pay	Demand Potential
On-line directory; yellow pages	384	9	Medium	High
Financial news	1,500	9	Medium	Low
Video phone	1,500	1,500	High	Low
Internet	3,000	384	High	High
Video conference	3,000	3,000	High	Low
Remote office	6,000	1,500	High	Medium
LAN interconnection	10,000	10,000	Medium	Medium
Supercomputing, CAD	45,000	45,000	High	Low

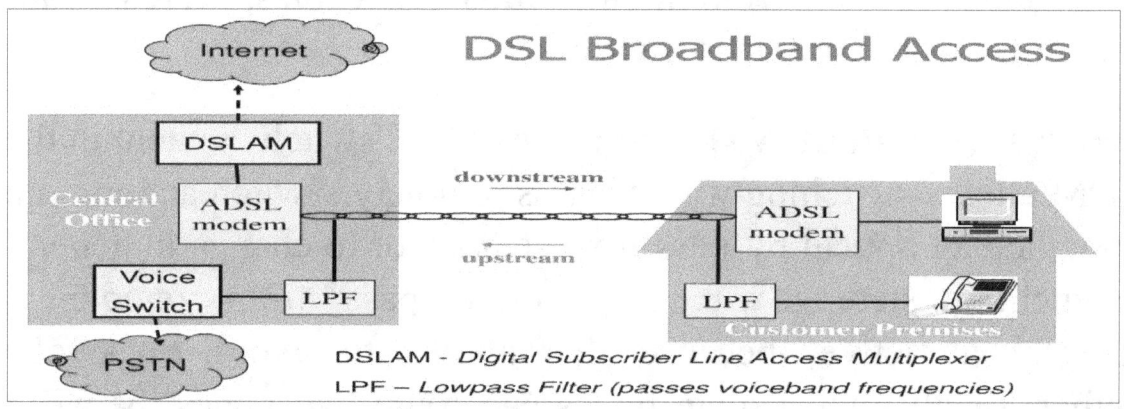

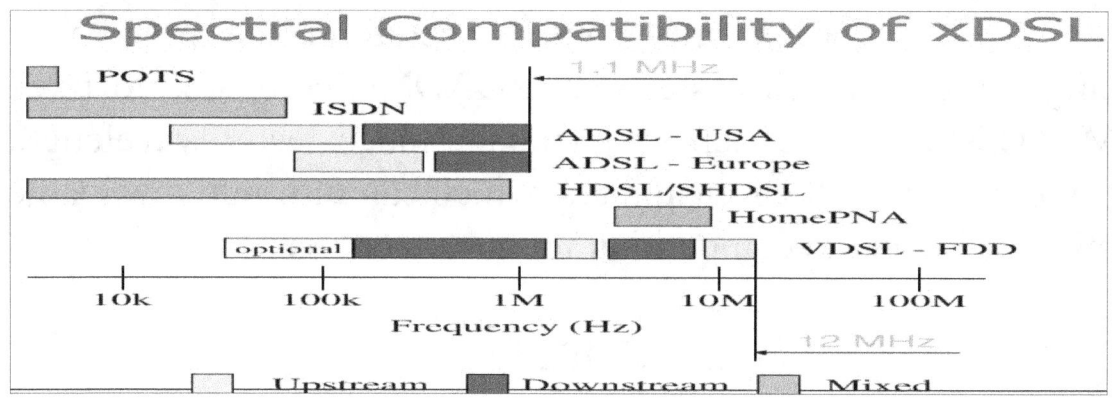

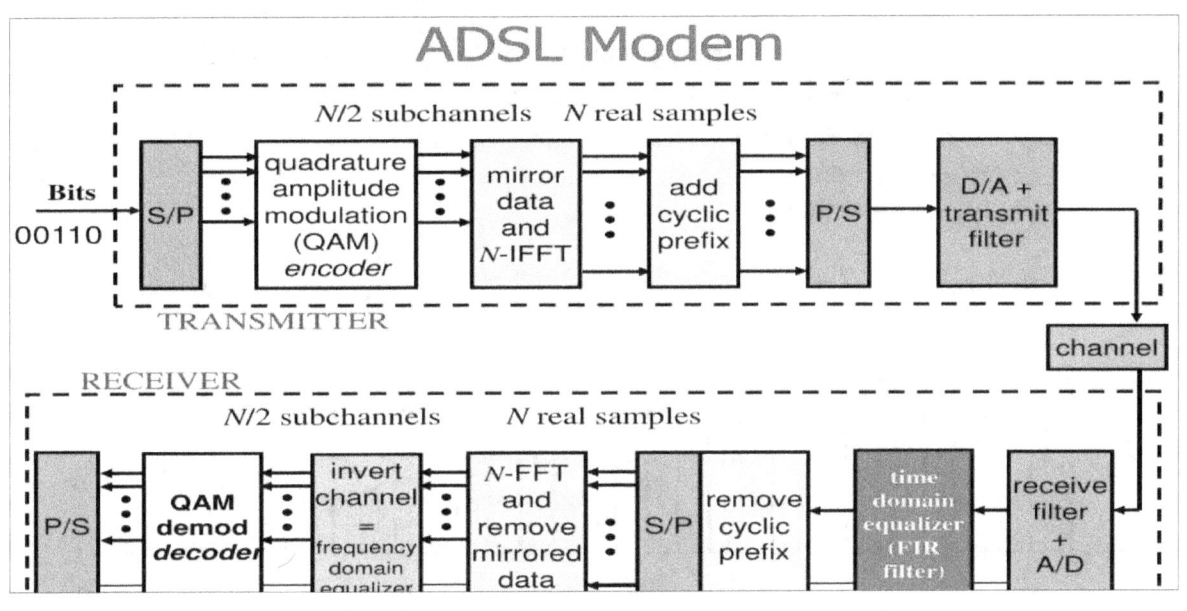

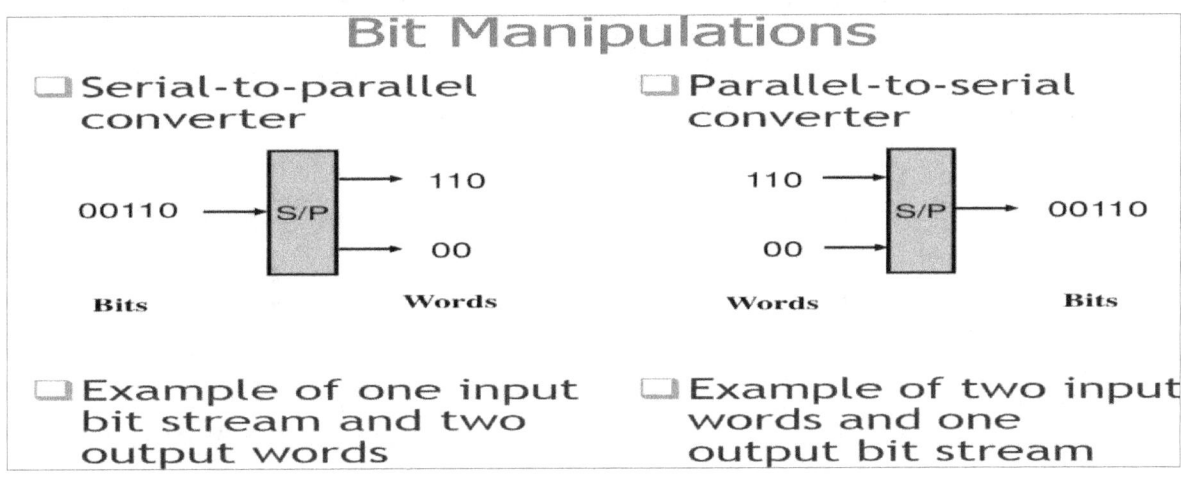

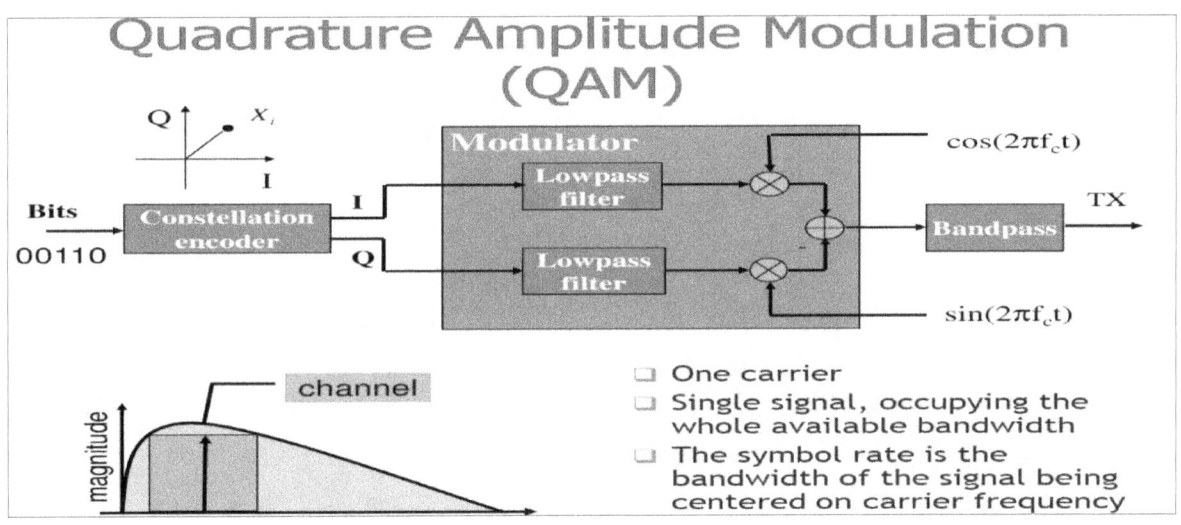

Multicarrier Modulation

- Divide broadband channel into narrowband subchannels
- Discrete Multitone (DMT) modulation
    - Based on fast Fourier transform (related to Fourier series)
    - Standardized for ADSL
    - Proposed for VDSL

Subchannels are 4.3 kHz wide in ADSL

every subchannel behaves like QAM

channel

carrier

subchannel

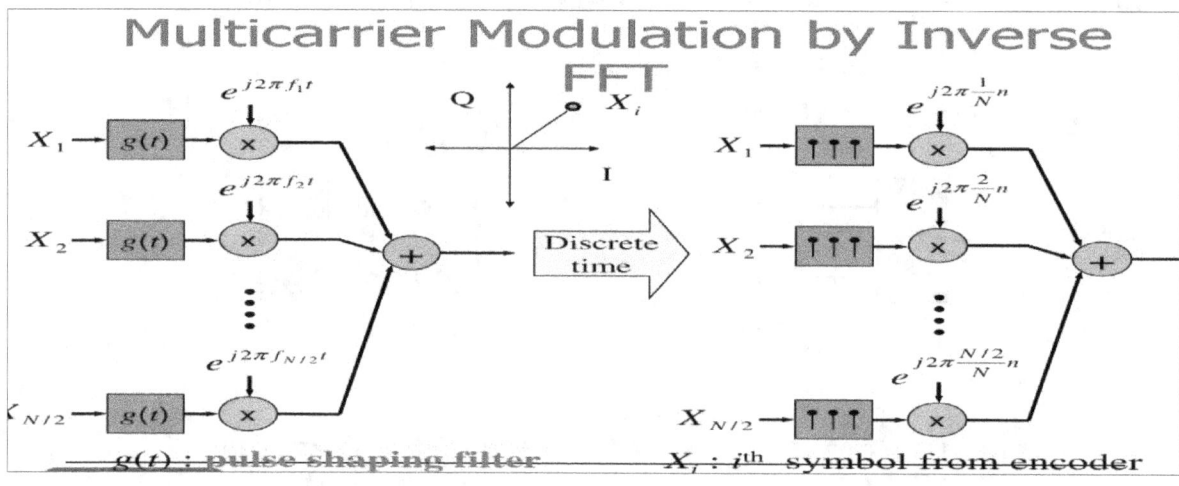

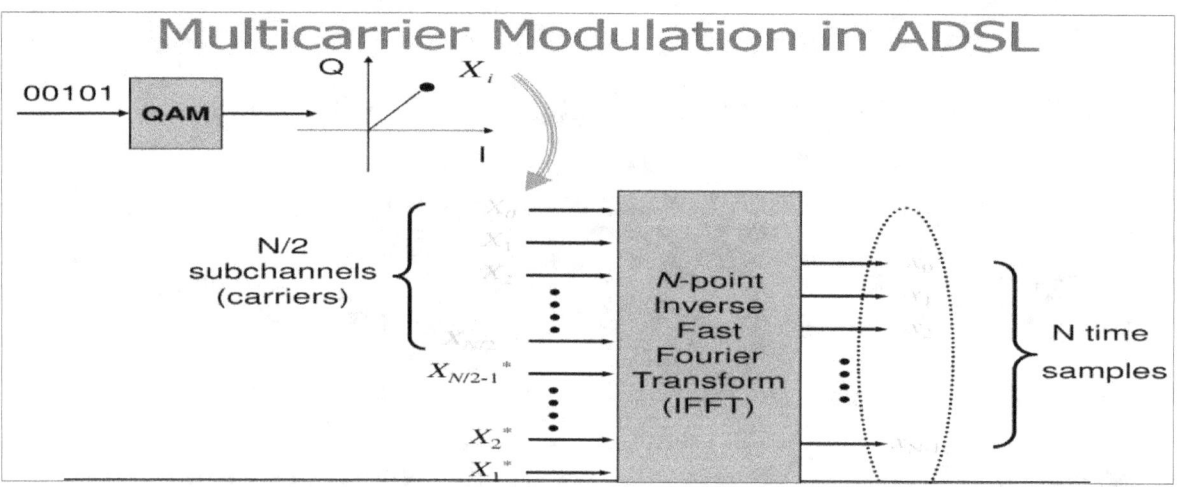

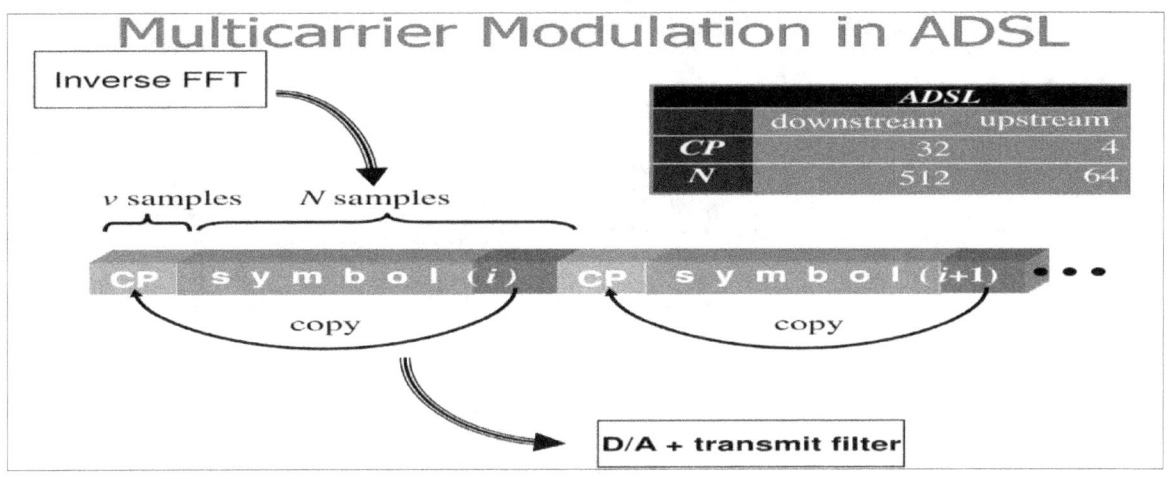

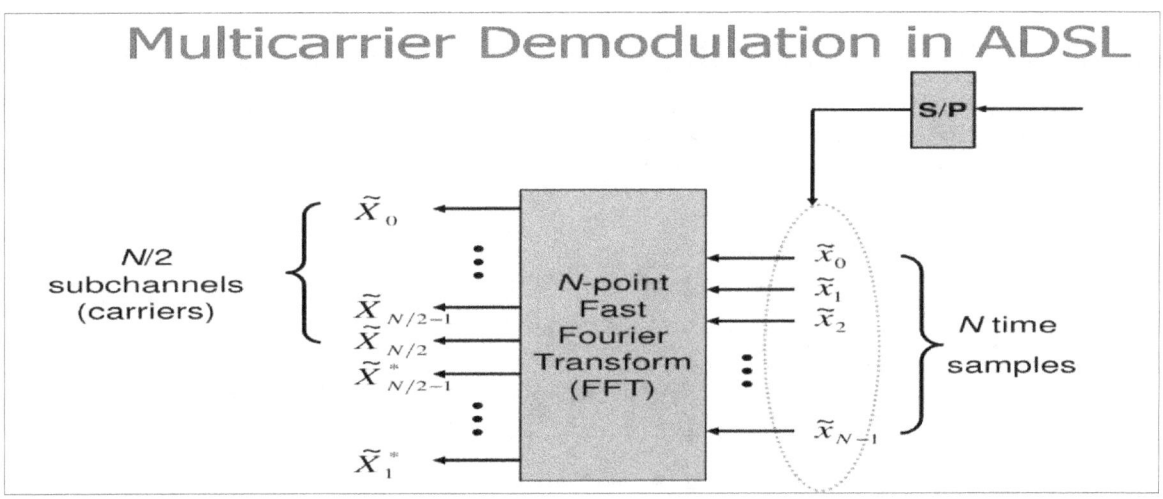

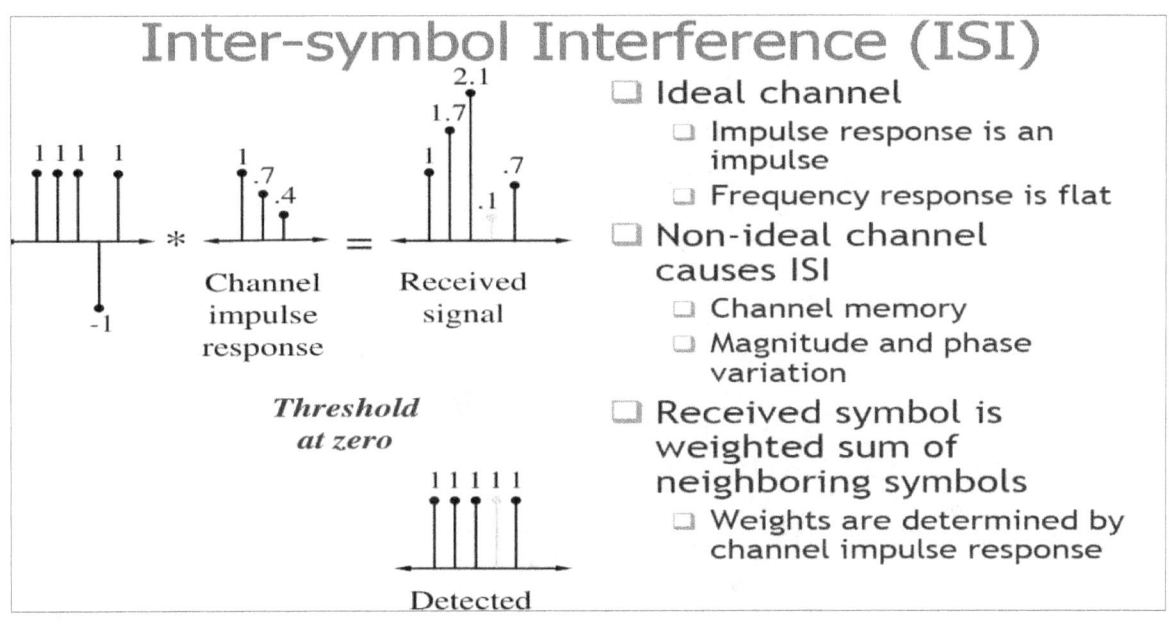

Channel Impulse Response

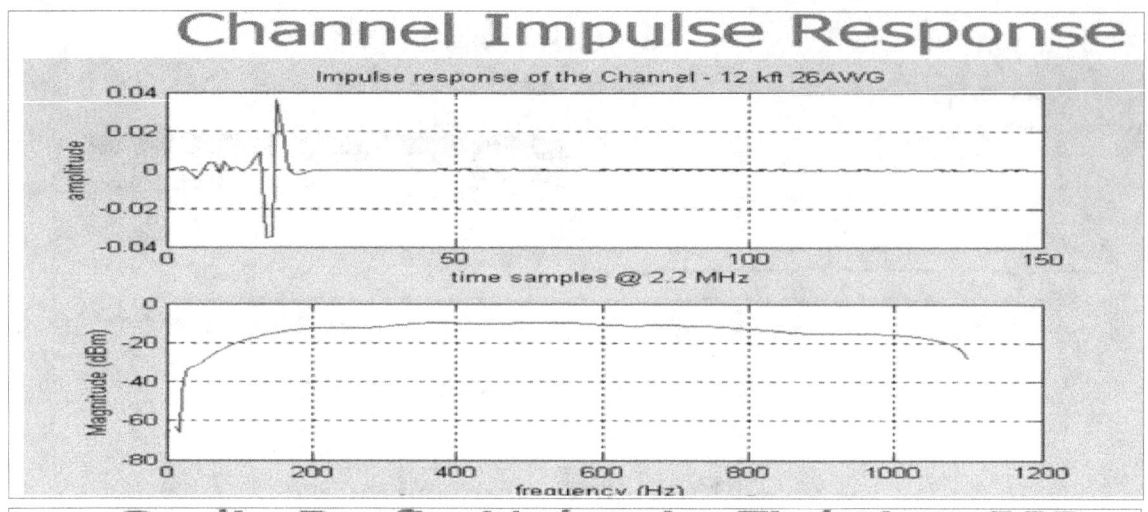

Cyclic Prefix Helps in Fighting ISI

- Provide guard time between successive symbols
  - No ISI if channel length is shorter than v +1 samples
- Choose guard time samples to be a copy of the beginning of the symbol - cyclic prefix
  - Cyclic prefix converts linear convolution into circular convolution
  - Need circular convolution so that
    symbol ⊗ channel ⇔ FFT(symbol) × FFT(channel)
  - Then division by the FFT(channel) can undo channel distortion

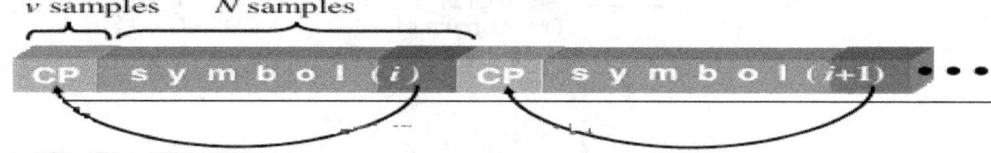

Combat ISI with Time-Domain Equalizer

- Channel length is usually longer than cyclic prefix
- Use finite impulse response (FIR) filter called a time-domain equalizer to shorten channel impulse response to be no longer than cyclic prefix length

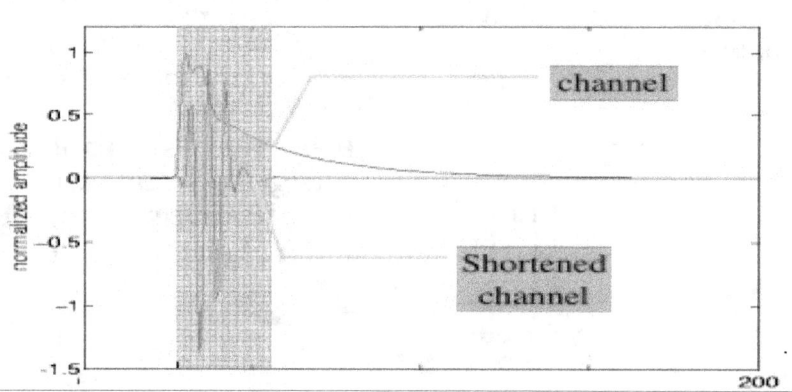

Convolution Review

- **Discrete-time convolution**

$$y[k] = \sum_{m=-\infty}^{\infty} h[m]\, x[k-m]$$

- **Continuous-time convolution**

$$y(t) = \int_{-\infty}^{\infty} h(\tau)\, x(t-\tau)\, d\tau$$

- For every k, we compute a new summation

  $x[k] \rightarrow \boxed{h[k]} \rightarrow y[k]$

  Represented by its impulse response

- For every value of t, we compute a new integral

  $x(t) \rightarrow \boxed{h(t)} \rightarrow y(t)$

  Represented by its impulse response

Finite Impulse Response (FIR) Filter

- Assuming that $h[k]$ is causal and has finite duration from $k = 0, \ldots, N-1$

$$y[k] = \sum_{m=0}^{N-1} h[m]\, x[k-m]$$

- Block diagram of an implementation (called a finite impulse response filter)

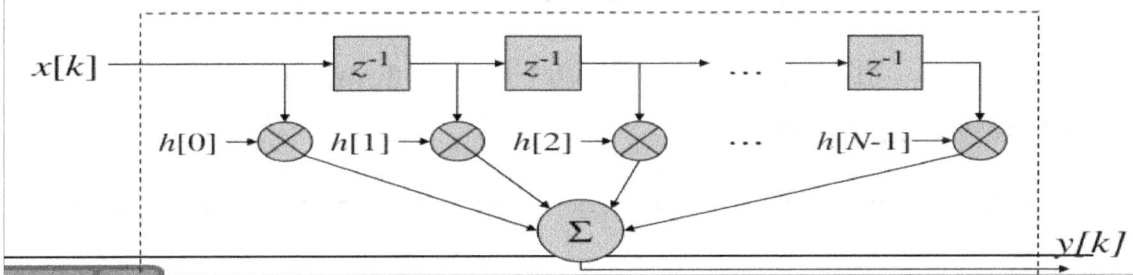

Frequency Domain Equalizer in ADSL

- Problem: FFT coefficients (constellation points) have been distorted by the channel.

- Solution: Use Frequency-domain Equalizer (FEQ) to invert the channel.

- Implementation: N/2 single-tap filters with complex coefficients.

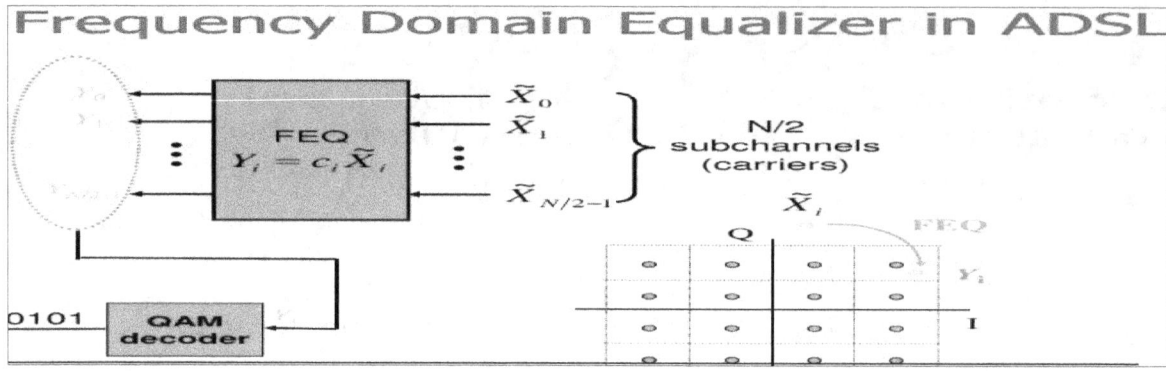

5.2 Radio Wave Propagation

The wireless radio channel puts fundamental limitations to the performance of wireless communications systems

Radio channels are extremely random, and are not easily analyzed

Modeling the radio channel is typically done in statistical fashion

Linear Path Loss

Suppose s(t) of power P_t is transmitted through a given channel

The received signal r(t) of power P_r is averaged over any random variations due to shadowing.

We define the linear path loss of the channel as the ratio of transmit power to receiver power

$$P_L = \frac{P_t}{P_r}$$

We define the path loss of the channel also in dB

$$P_L \text{ dB} = 10 \log_{10} \frac{P_t}{P_r} \text{ dB (nonnegative number)}$$

Experimental results

The measurements and predictions for the receiving van driven along 19th St./Nash St.

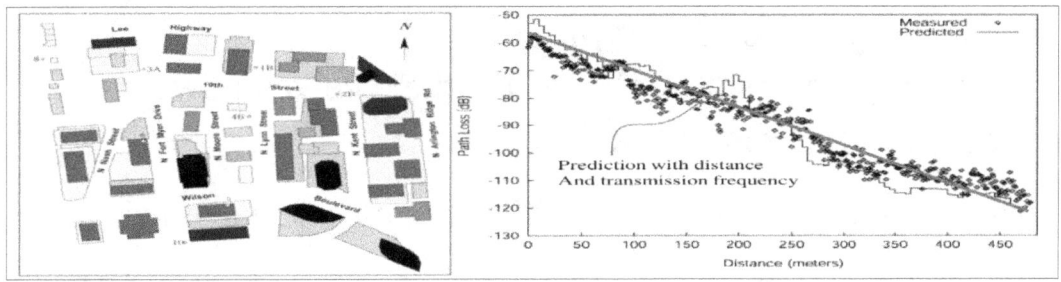

Line-of-Sight Propagation

Attenuation

The strength of a signal falls off with distance

Free Space Propagation

The transmitter and receiver have a clear line of sight path between them. No other sources of impairment!

Satellite systems and microwave systems undergo free space propagation

The free space power received by an antenna which is separated from a radiating antenna by a distance is given by Friis free space equation

Friis Free Space Equation

The relation between the transmit and receive power is given by Friis free space equations:

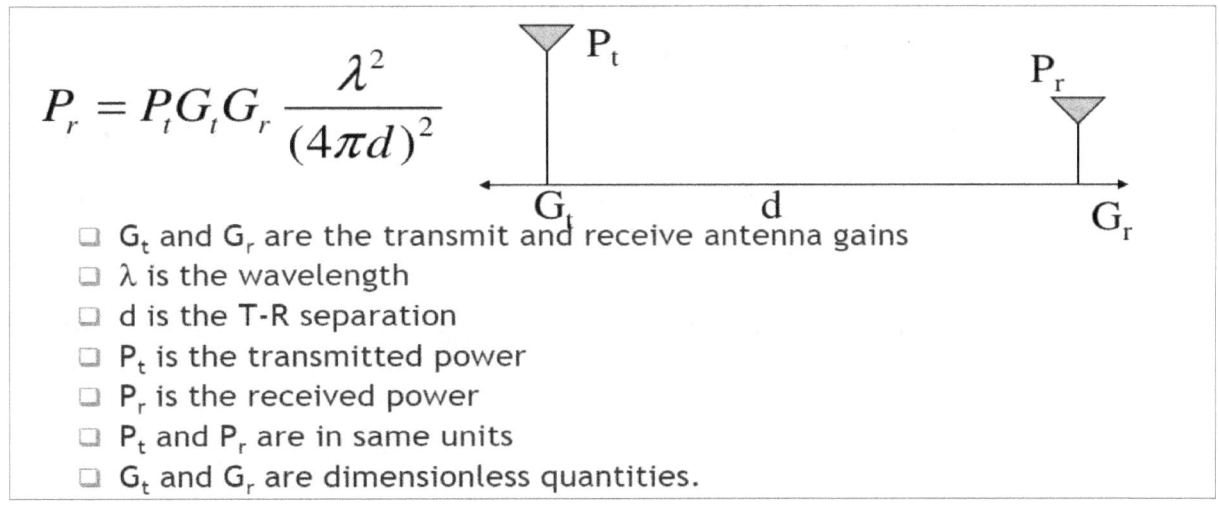

$$P_r = P_t G_t G_r \frac{\lambda^2}{(4\pi d)^2}$$

- G_t and G_r are the transmit and receive antenna gains
- λ is the wavelength
- d is the T-R separation
- P_t is the transmitted power
- P_r is the received power
- P_t and P_r are in same units
- G_t and G_r are dimensionless quantities.

Free Space Propagation Example:

The Friis free space equation shows that the received power fall off as the square of the T-R separation distances.

- The received power decays with distance by 20 dB/decade
  - EX: Determine the isotropic free space loss at 4 GHz for the shortest path to a geosynchronous satellite from earth (35,863 km).
  $P_L = 20\log_{10}(4\times10^9) + 20\log_{10}(35.863\times10^6) - 147.56 dB$
  $P_L = 195.6$ dB

  Suppose that the antenna gain of both the satellite and groundbased antennas are 44 dB and 48 dB, respectively

  $P_L = 195.6 - 44 - 48 = 103.6$ Db

  - Now, assume a transmit power of 250 W at the earth station. What is the power received at the satellite antenna?

Basic Propagation Mechanisms

Reflection, diffraction, and scattering:

Reflection occurs when a propagating electromagnetic wave impinges upon an object

Diffraction occurs when the radio path between the transmitter and receiver is obstructed by a surface that has sharp edges

Scattering occurs when the medium through which the wave travels

consists of objects with dimensions that are small compared to the wavelength, or

the number of obstacles per unit volume is large.

Basic Propagation Mechanisms

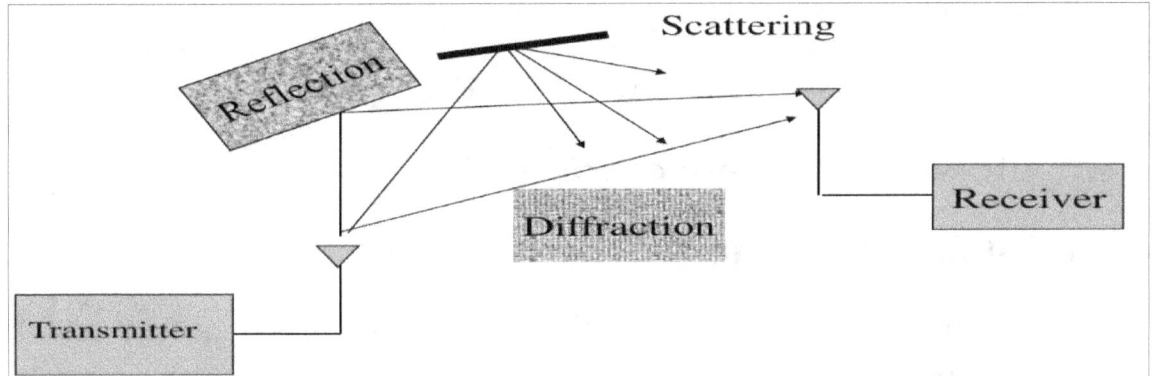

Free Space Propagation

Can be also expressed in relation to a reference point, d0

$$P_r(d) = P_t K \left(\frac{d_o}{d}\right)^2 \qquad d \geq d_o$$

K is a unitless constant that depends on the antenna characteristics and free-space path loss up to distance d0

- Typical value for d_o:
  - Indoor: 1m
  - Outdoor: 100m to 1 km

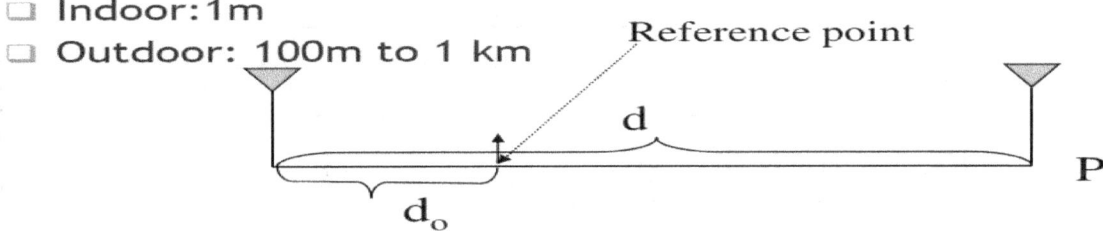

Simplified Path Loss Model

Complex analytical models or empirical measurements when tight system specifications must be met

Best locations for base stations

Access point layouts

However, use a simple model for general tradeoff analysis

$$P_r = P_t K \left[\frac{d_o}{d}\right]^\gamma$$

- dB attenuation model

$$P_r \text{ dBm} = P_t \text{ dBm} + K \text{ dB} - 10\gamma \log_{10}\left[\frac{d}{d_0}\right]$$

- d_0: close-in reference point

Typical Pathloss Exponents

Empirically, the relation between the average received power and the distance is determined by the expression where γ is called the path loss exponent

- The typical values of γ are as: $P_r \propto d^{-\gamma}$

Environment	Path Loss exponent, γ
Free Space	2
Urban Area	2.7 to 3.5
Suburban Area	3 to 5
Indoor (line-of-sight)	1.6 to 1.8

Radio System Design

Fade Margins: The difference between the normal received power and the power required for minimum acceptable performance is referred to as the fade margin. Greater fade margins imply less frequent occurrences of minimum performance levels.

When large fade margins are provided, the received signal power during unfaded conditions is so strong that bit errors are virtually nonexistent.

To minimize dynamic range requirements in a receiver and reduce interference between systems, adaptive transmitter power control

(ATPC) is sometimes used. Thus, when excess power is unnecessary, it is not used.

Noise Power

Noise power in a receiver is usually dominated by thermal noise generated in the frontend receiver amplifier. In this case, the noise power can be determined as follows

$$P_N = FkT_0B$$

F=the receiver noise figure
T_0= the reference receiver temperature in degrees Kelvin (290⁰)
K=1.38× 10⁻²³ is Boltzmann's constant
B=the receiver bandwidth

- The noise figure of any device is defined as the ratio of the input SNR to the output SNR.
$$F=SNR_{in}/SNR_{out}$$

System Gain/Fade Margin

System Gain is defined to be the difference, in decibels, of the transmitter output power and the minimum receive power for the specified error rate:

$$A_s = 10 \log_{10}\left(\frac{P_t}{P_{req}}\right)$$

- Combining the noise figure and the system gain equations:
$$A_s = 10\log_{10}\left(\frac{P_t}{SNR F k T_0 B}\right) - D$$

  D is the degradation from the ideal performance
  SNR=P_{req}/P_N
- System gain, in conjunction with antenna gains and path losses, determines the fade margin (assuming free space path loss)
  Fade Margin = $A_s + G_T + G_R + 20\log_{10}\lambda - A_f - 20\log_{10}(4\pi d)$
  - A_f=system (branching and coupling) loss, G_T and G_R=transmit and receive antenna gains, λ=transmitted wavelength(λ=c/f_c), d=distance

Cell Radious Prediction:

- The signal level is same on a circle centered at the base station with radius R
- Find the distance R such that the received signal power cannot be less than P_{min} dBm
- The received signal power at a distance d=R is specified by

$$P_r(d)(dB) = P_t(dB) + 10\log_{10} K - 10\gamma \log_{10}\left(\frac{d}{d_0}\right)$$

$$P_r(R) \leq P_{min}$$

- Solving the above equation for the radius R, we obtain

$$R \leq d_0 \times 10^{0.1(P_T/\gamma)}$$

  - where $P_T = P_{min} - P_t - 10\log_{10} K$

Mobil Network:

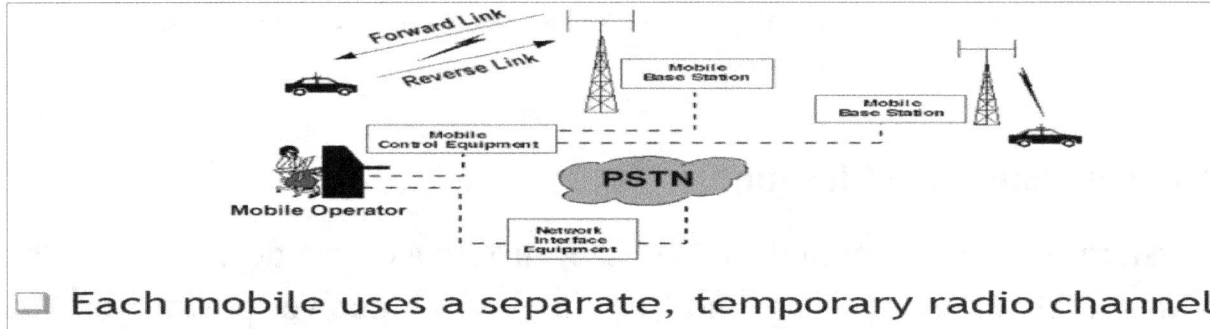

- Each mobile uses a separate, temporary radio channel
- The cell site talks to many mobiles at once
- Channels use a pair of frequencies for communication
  - forward link
  - reverse link

Early Mobile Telephone System:

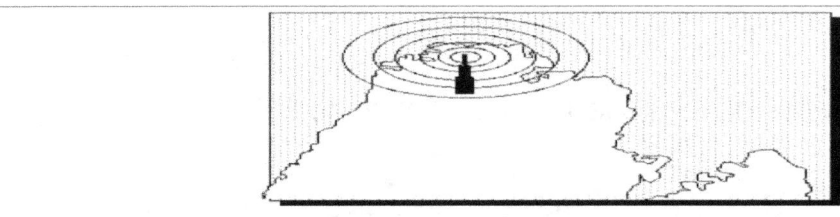

- Traditional mobile service was structured in a fashion similar to television broadcasting
- One very powerful transmitter located at the highest spot in an area would broadcast in a radius of up to 50 kilometers
- This approach achieved very good coverage, but it was impossible to reuse the frequencies throughout the system because of interference

Cellular Approach

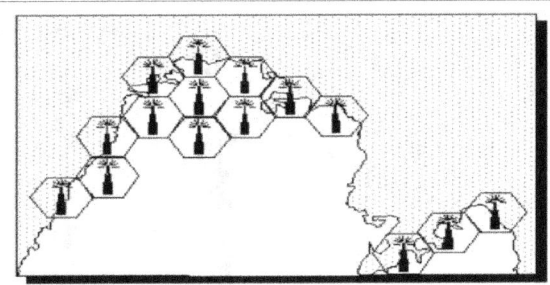

- Instead of using one powerful transmitter, many low-power transmitters were placed throughout a coverage area to increase the capacity
- Each base station is allocated a portion of the total number of channels available to the entire system
- To minimize interference, neighboring base stations are assigned different groups of channels

Why Cellular:

- By systematically spacing base stations and their channel groups, the available channels are:
  - distributed throughout the geographic region
  - maybe reused as many times as necessary provided that the interference level is acceptable
- As the demand for service increases the number of base stations may be increased thereby providing additional radio capacity
- This enables a fixed number of channels to serve an arbitrarily large number of subscribers by reusing the channel throughout the coverage region

Cells:

- A cell is the basic geographic unit of a cellular system
- The term *cellular* comes from the honeycomb shape of the areas into which a coverage region is divided
- Each cell size varies depending on the landscape
- Because of constraints imposed by natural terrain and man-made structures, the true shape of cells is not a perfect hexagon

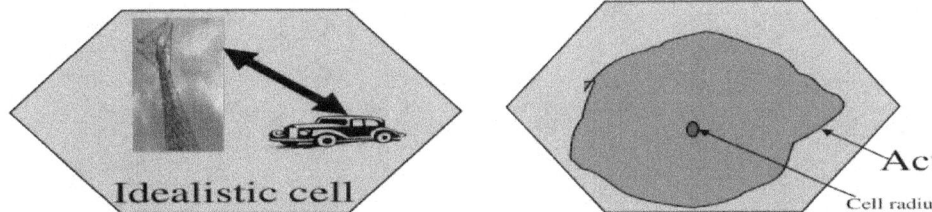

Cell Cluster Concept:

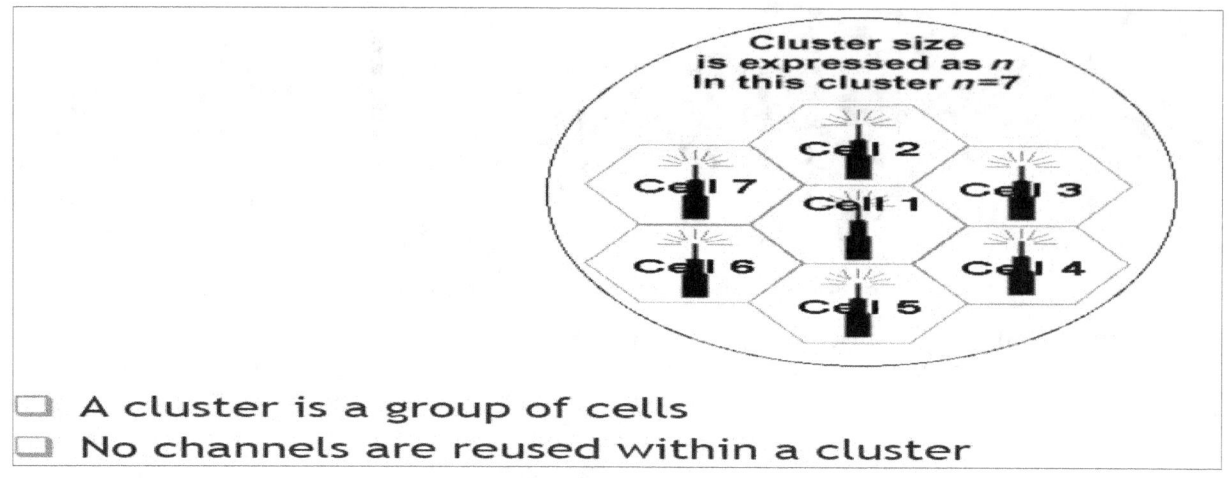

- A cluster is a group of cells
- No channels are reused within a cluster

Frequency Reuse:

- Cells with the same number have the same set of frequencies
- 3 clusters are shown in the figure
- Cluster size N = 7
- Each cell uses 1/N of available cellular channels (frequency reuse factor)

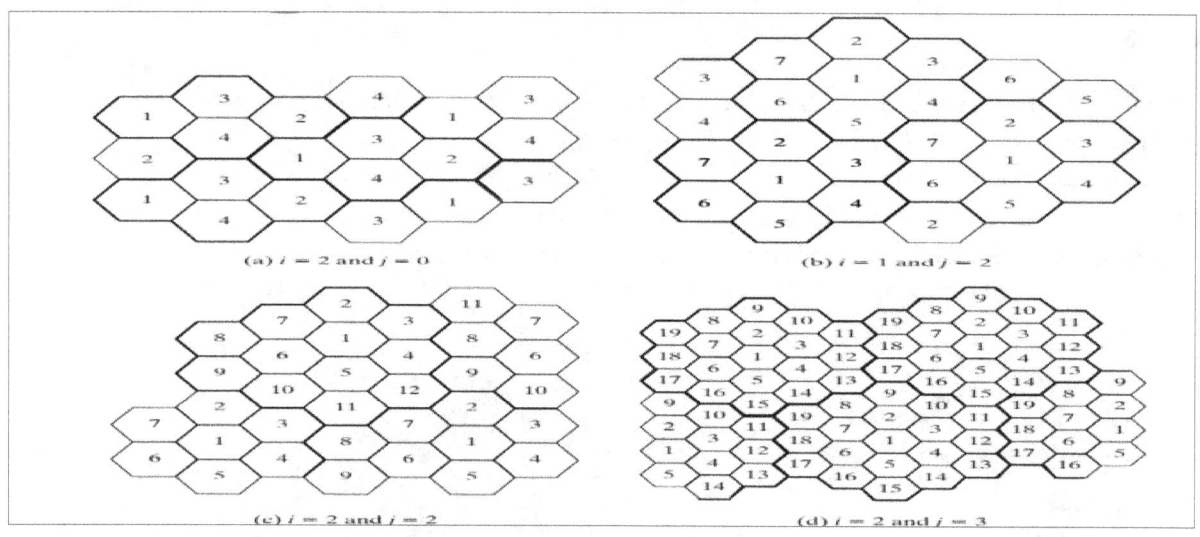

Methods for finding Co-channel Cells:

(a) $i = 2$ and $j = 0$

(b) $i = 1$ and $j = 2$

(c) $i = 2$ and $j = 2$

(d) $i = 2$ and $j = 3$

Geometry of Hexagonal cells:

- Distance between nearest cochannel cells
- A hexagon has exactly six equidistant neighbors separated by multiple of 60 degrees
- Approximate distance between the centers of two nearest cochannel cells is

$$D = \sqrt{3N}\,R$$

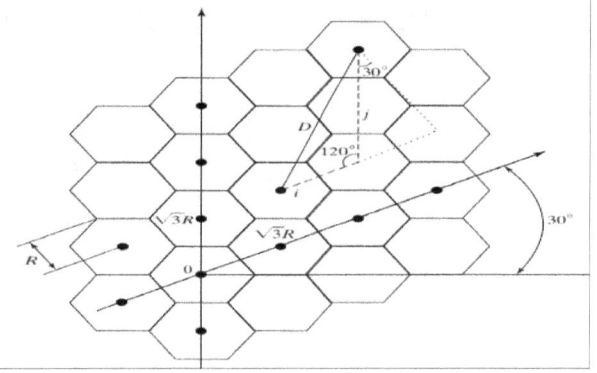

Frequency Reuse Ratio:

Signal

- The frequency reuse ratio is defined as

$$q = \frac{D}{R}$$

- The frequency reuse patterns below apply to hexagonal cells,

$$q = \frac{D}{R} = \sqrt{3N}$$

Frequency Reuse Pattern (i, j)	Cluster Size N	Frequency Reuse Ratio q
(1, 1)	3	3.00
(2, 0)	4	3.46
(2, 1)	7	4.58
(3, 0)	9	5.20
(2, 2)	12	6.00
(3, 1)	13	6.24
(3, 2)	19	7.55
(4, 1)	21	7.94
(3, 3)	27	9.00
(4, 2)	28	9.17
(4, 3)	37	10.54

Signal to Interference ratio

- Let N_I be the number of co-channel interfering cells
- P_r is the desired signal power from the desired base station
- P_i is the interference power caused by the ith interfering co-channel cell base station
- The SIR (S/I) at the desired mobile receiver is

$$\frac{S}{I} = \frac{P_r}{\sum_{i=1}^{N_I} P_i}$$

Recall Power-Distance Relation:

- Average received signal strength at any point in a mobile radio channel is

$$P_r = P_t K \left(\frac{d}{d_0}\right)^{-\gamma}$$

  - If d_0 is the close-in reference point in the far field region of the antenna from the transmitting antenna
  - P_t is the transmitter power
  - γ is the path loss exponent
  - P_r is the received power at a distance d

Approximated SIR:

- SIR for a mobile can be approximated as

$$\frac{S}{I} = \frac{R^{-\gamma}}{\sum_{i=1}^{N_I}(D_i)^{-\gamma}}$$

  - If the transmit power of each base station is equal
  - γ is same throughout the coverage area
  - D_i is the distance of the ith interferer from the mobile
- SIR as considering only the first layer of interfering cells can be simplified as

$$\frac{S}{I} = \frac{(D/R)^\gamma}{N_I} = \frac{(\sqrt{3N})^\gamma}{N_I}$$

  - if all interfering base stations are equi-distant from each other and this distance is $D_i \approx D$

- With hexagon shaped cellular systems, there are always six cochannel interfering cells in the first tier.
- The frequency reuse ratio can be expressed as

$$q = \left(N_I \times \frac{S}{I}\right)^{1/\gamma} = \left(6 \times \frac{S}{I}\right)^{1/\gamma}.$$

- Example: For the U.S. AMPS analog FM system, a value of S/I = 18 dB or greater is acceptable.
  - With a path loss exponent of $\gamma=4$, the frequency reuse ratio q is determined as

$$q = (6 \times 10^{1.8})^{1/4} = (6 \times 63.1)^{0.25} \simeq 4.41.$$

  - Therefore, the cluster size N should be

$$N = q^2/3 = 6.49 \simeq 7.$$

S/I Ratio vs Cluster Size:

❏ Suppose the acceptable S/I in a cellular system is 20 dB. γ=4, what is the minimum cluster size? Consider only the closest interferers.

5.2 GPON –Fundamentals

What is PON?

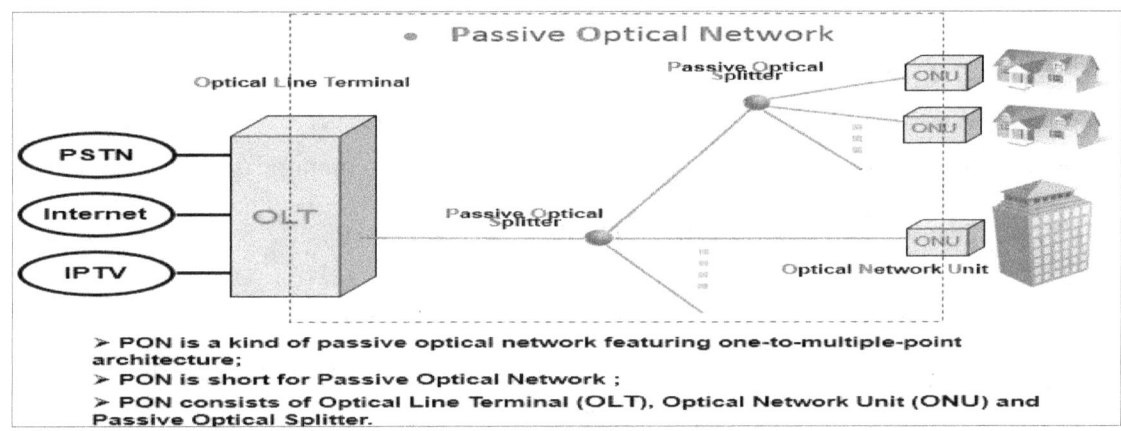

> PON is a kind of passive optical network featuring one-to-multiple-point architecture;
> PON is short for Passive Optical Network ;
> PON consists of Optical Line Terminal (OLT), Optical Network Unit (ONU) and Passive Optical Splitter.

Why GPON?

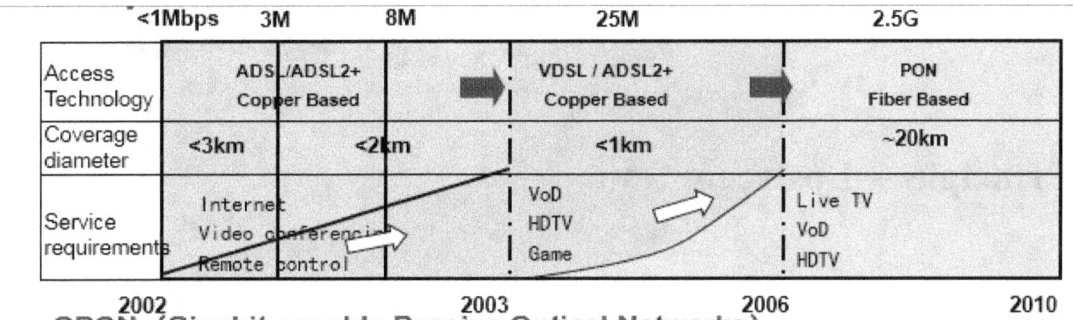

- GPON (Gigabit-capable Passive Optical Networks)
- GPON supports Triple – play service, providing competitive all-service solution.
- GPON supports high-bandwidth transmission to break down the bandwidth bottleneck of the access over twisted pair cables, so as to satisfy the requirements of high-bandwidth services, such as IPTV and live TV broadcasts.
- GPON supports the long-reach (up to 20 km) service coverage to overcome the obstacle of the access technology over twisted pair cables and reduce the network nodes.
- With complete standards and high technical requirements, GPON supports integrated services in a good way.
- GPON is the choice of large carriers in the international market.

GPON Principle----Data Multiplexing

- GPON adopts Wavelength Division Multiplexing (WDM) technology, facilitating bi-direction communication over a single fiber.

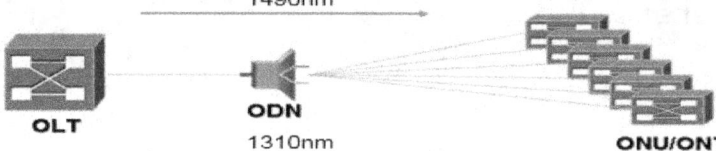

- To separate upstream/downstream signals of multiple users over a single fiber, GPON adopts
- two multiplexing mechanism:
  - In downstream direction, data packets are transmitted in a broadcast manner;
  - In upstream direction, data packets are transmitted in a TDMA manner.

GPON Principle----Downstream Data

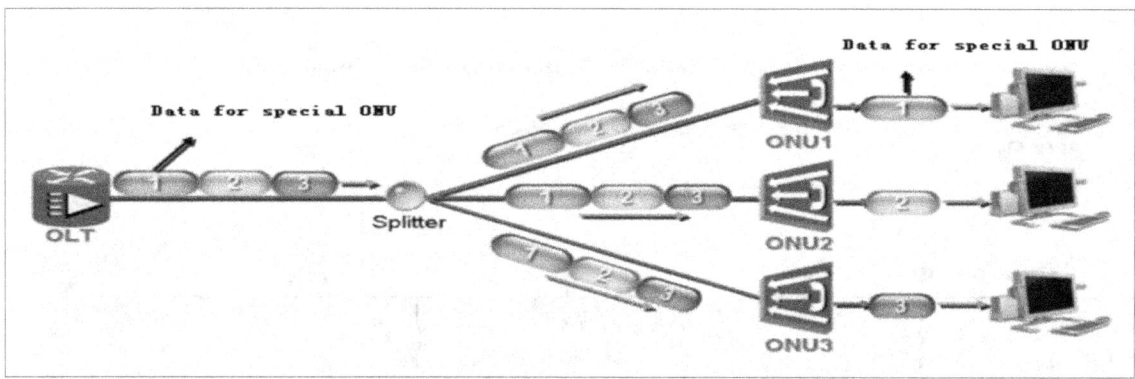

GPON Principle----Upstream Data

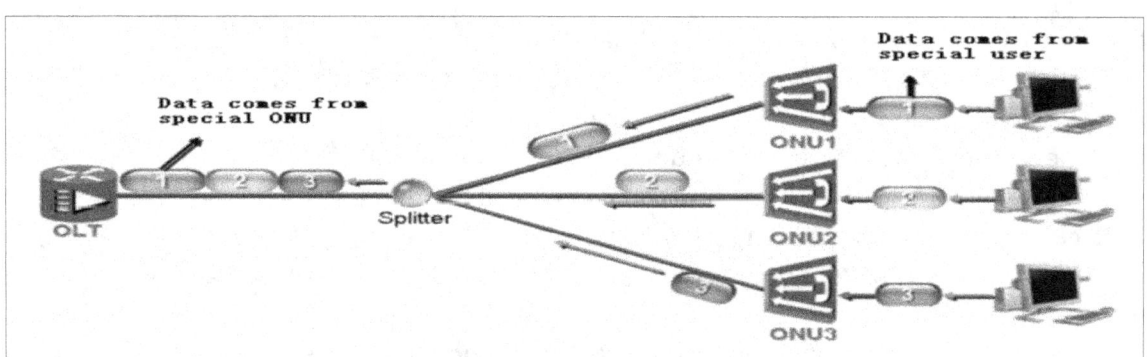

What is Optical Access Network?

From the architecture diagram, the optical access network comprises the following scenarios:

1. FTTB scenario

SBU: Single business unit; providing a comparatively small number of ports such as POTS, 10/100/1000BASE-T, RF, and DS1/T1/E1 ports

MTU: Business Multi-tenant unit; providing a comparatively larger number of ports, including POTS, 10/100/1000BASE-T, RF and DS1/T1/E1 ports.

2. FTTC & FTT Cab scenario

MDU : Multi-dwelling unit ;providing a comparatively larger number of ports, including 10/100/1000BASE-T, RF, VDSL2, and so on.

3. FTTH scenario

SFU : Single family unit , providing a comparatively small number of ports, including following types: POTS, 10/100/1000BASE-T, and RF.

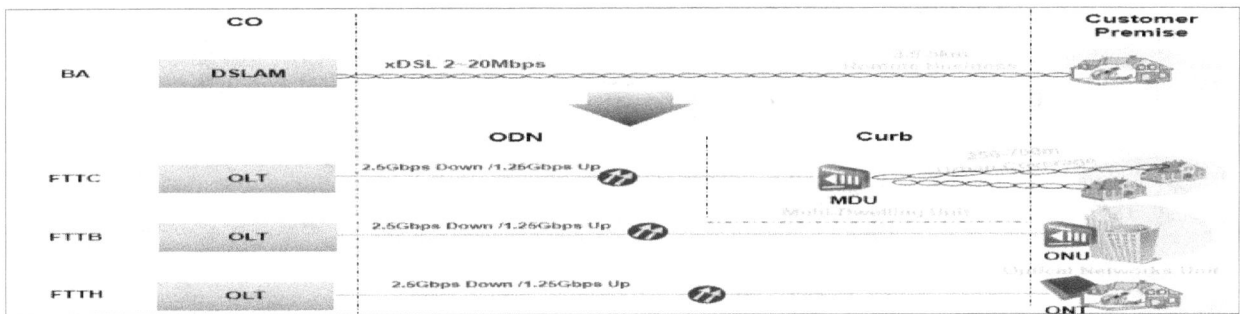

GPON Standards

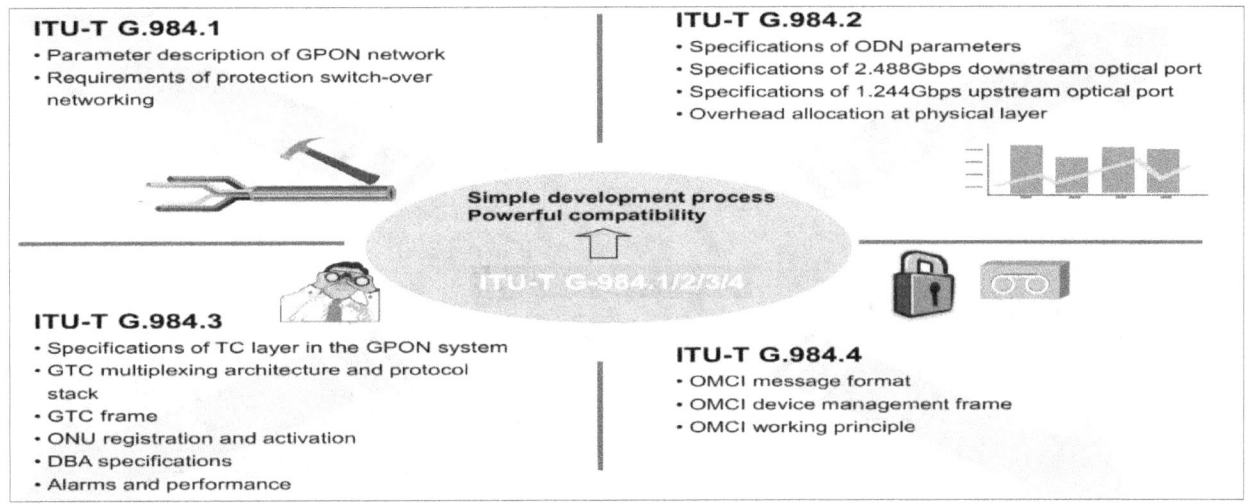

GPON Network Model Reference

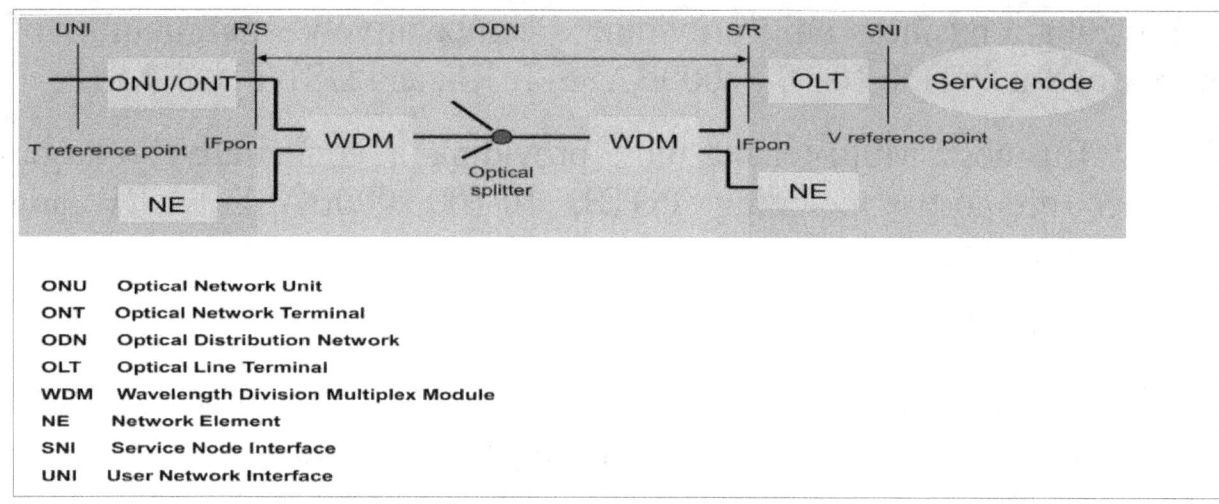

- ONU — Optical Network Unit
- ONT — Optical Network Terminal
- ODN — Optical Distribution Network
- OLT — Optical Line Terminal
- WDM — Wavelength Division Multiplex Module
- NE — Network Element
- SNI — Service Node Interface
- UNI — User Network Interface

Basic Performance Parameters

- GPON identifies 7 transmission speed combination as follows:
  - 0.15552 Gbit/s up, 1.24416 Gbit/s down
  - 0.62208 Gbit/s up, 1.24416 Gbit/s down
  - 1.24416 Gbit/s up, 1.24416 Gbit/s down
  - 0.15552 Gbit/s up, 2.48832 Gbit/s down
  - 0.62208 Gbit/s up, 2.48832 Gbit/s down
  - 1.24416 Gbit/s up, 2.48832 Gbit/s down
  - 2.48832 Gbit/s up, 2.48832 Gbit/s down

  Among them, 1.24416 Gbit/s up, 2.48832 Gbit/s down is the mainstream speed combination supported at current time.

- Maximum logical reach: 60 km
- Maximum physical reach: 20 km
- Maximum differential fibre distance: 20 km
- Split ratio: 1 : 64, it can be up to 1 : 128

GPON Multiplexing Architecture

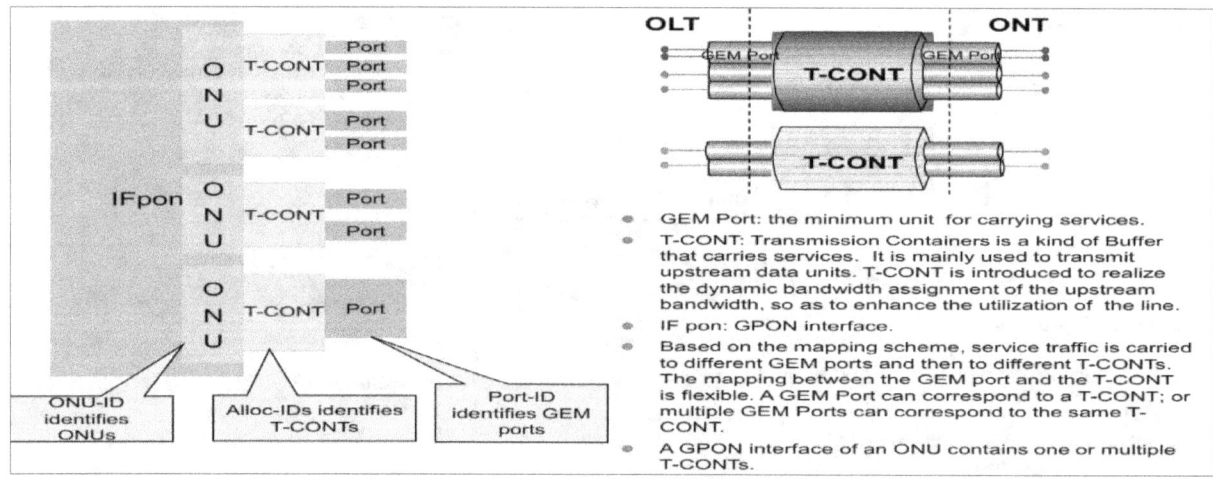

- GEM Port: the minimum unit for carrying services.
- T-CONT: Transmission Containers is a kind of Buffer that carries services. It is mainly used to transmit upstream data units. T-CONT is introduced to realize the dynamic bandwidth assignment of the upstream bandwidth, so as to enhance the utilization of the line.
- IF pon: GPON interface.
- Based on the mapping scheme, service traffic is carried to different GEM ports and then to different T-CONTs. The mapping between the GEM port and the T-CONT is flexible. A GEM Port can correspond to a T-CONT; or multiple GEM Ports can correspond to the same T-CONT.
- A GPON interface of an ONU contains one or multiple T-CONTs.

GPON Frame Structure

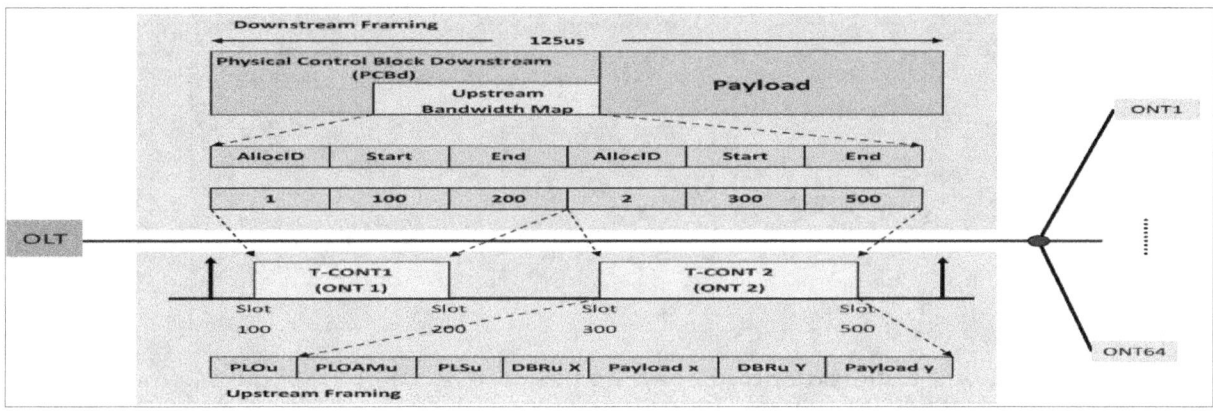

GPON Upstream Frame Structure

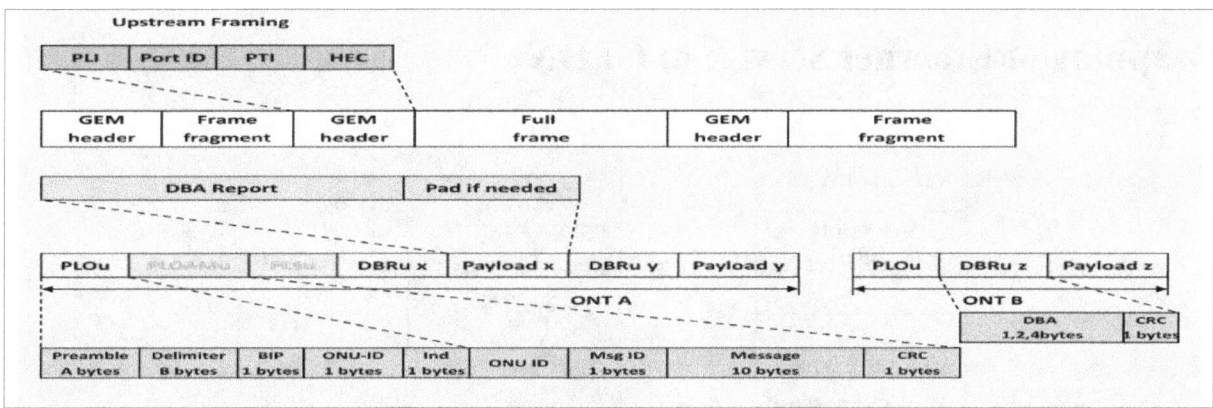

GPON Downstream Frame Structure

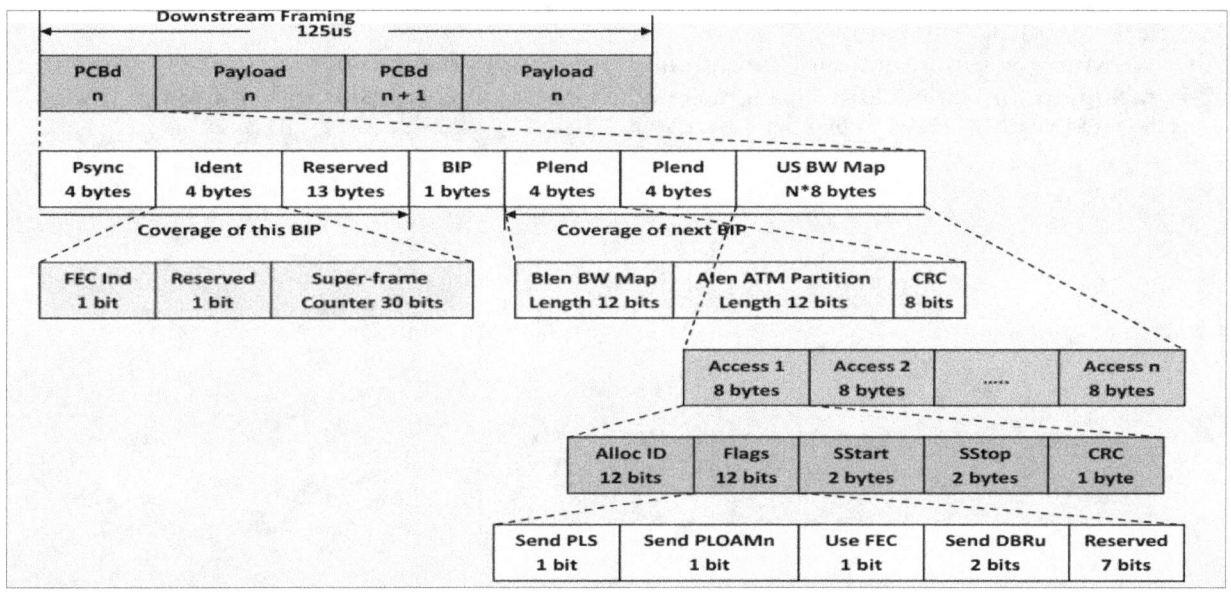

Mapping of TDM Service in GPON

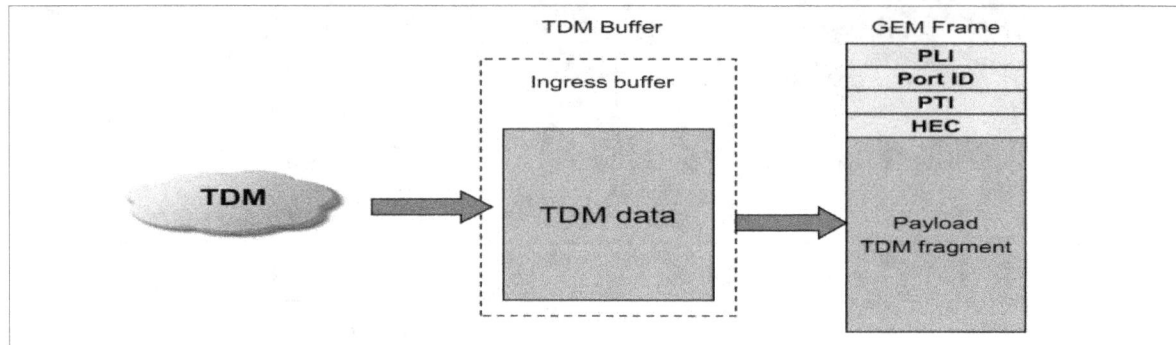

- TDM frames are buffered and queued as they arrive, then TDM data is multiplexed in to fixed-length GEM frames for transmission.
- This scheme does not vary TDM services but transmit TDM services transparently.
- Featuring fixed length, GEM frames benefits the transmission of TDM services.

Mapping of Ethernet Service in GPON

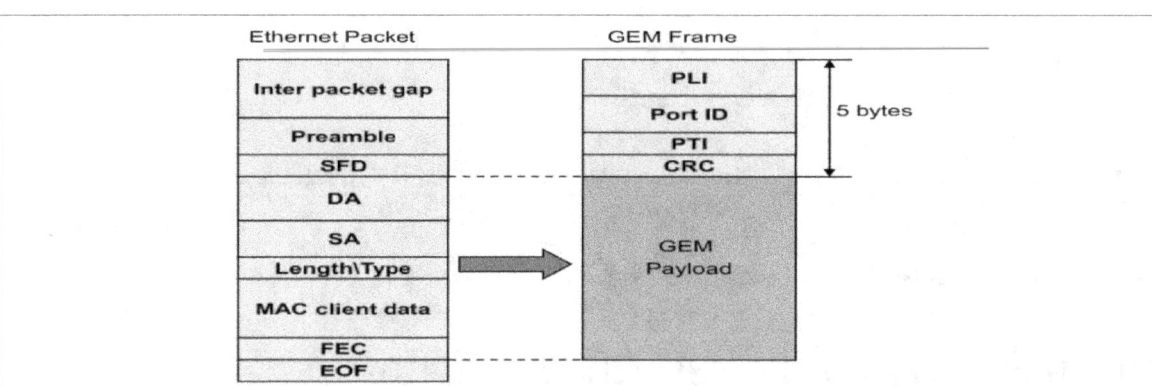

- GPON system resolves Ethernet frames and then directly maps the data of frames into the GEM Payload.
- GEM frames automatically encapsulate header information.
- Mapping format is clear and it is easy for devices to support this mapping. It also boasts good compatibility.

References

[1] ITU-T 984.1, ITU-T 984.2, ITU-T 984.3, ITU-T 984.4

[2] Understanding Telecommunications Networks, Andy Valdar, 2006 The Institution of Engineering and Technology

[3] Digital Telephony (John Bellamy) and Fundamentals of Telecommunications (R. L. Freeman)

[4] Introduction to Telecommunications Network Engineering (Second Edition), (Anttalainen, Tarmo)

[5] Fundamentals of PCM, 02_tt2550eu01al_01,02,03

[6] WDM technology siemens course

www.ingramcontent.com/pod-product-compliance
Lightning Source LLC
Chambersburg PA
CBHW060411220526
45465CB00008B/2840